INNOVATION POLICIES FOR THE 21ST CENTURY

Report of a Symposium

Committee on Comparative Innovation Policy:
Best Practice for the 21st Century

Board on Science, Technology, and Economic Policy

Policy and Global Affairs

Charles W. Wessner, Editor

NATIONAL RESEARCH COUNCIL
OF THE NATIONAL ACADEMIES

THE NATIONAL ACADEMIES PRESS
Washington, D.C.
www.nap.edu

THE NATIONAL ACADEMIES PRESS 500 Fifth Street, N.W. Washington, DC 20001

NOTICE: The project that is the subject of this report was approved by the Governing Board of the National Research Council, whose members are drawn from the councils of the National Academy of Sciences, the National Academy of Engineering, and the Institute of Medicine. The members of the committee responsible for the report were chosen for their special competences and with regard for appropriate balance.

This study was supported by: Contract/Grant No. SB1341-03-C-0032 between the National Academy of Sciences and the U.S. Department of Commerce; Contract/Grant No. OFED-381989 between the National Academy of Sciences and Sandia National Laboratories; and Contract/Grant No. NAVY-N00014-05-G-0288, DO #2, between the National Academy of Sciences and the Office of Naval Research. This material is based upon work also supported by the Defense Advanced Research Projects Agency Defense Sciences Office, DARPA Order No. K885/00, Program Title: Materials Research and Development Studies, Issued by DARPA/CMD under Contract #MDA972-01-D-0001. Additional funding was provided by Intel Corporation, International Business Machines, and Google. Any opinions, findings, conclusions, or recommendations expressed in this publication are those of the author(s) and do not necessarily reflect the views of the organizations or agencies that provided support for the project.

International Standard Book Number-13:   978-0-309-10316-9
International Standard Book Number-10:   0-309-10316-9

Limited copies are available from Board on Science, Technology, and Economic Policy, National Research Council, 500 Fifth Street, N.W., W547, Washington, DC 20001; 202-334-2200.

Additional copies of this report are available from the National Academies Press, 500 Fifth Street, N.W., Lockbox 285, Washington, DC 20055; (800) 624-6242 or (202) 334-3313 (in the Washington metropolitan area); Internet, http://www.nap.edu.

THE NATIONAL ACADEMIES
Advisers to the Nation on Science, Engineering, and Medicine

The **National Academy of Sciences** is a private, nonprofit, self-perpetuating society of distinguished scholars engaged in scientific and engineering research, dedicated to the furtherance of science and technology and to their use for the general welfare. Upon the authority of the charter granted to it by the Congress in 1863, the Academy has a mandate that requires it to advise the federal government on scientific and technical matters. Dr. Ralph J. Cicerone is president of the National Academy of Sciences.

The **National Academy of Engineering** was established in 1964, under the charter of the National Academy of Sciences, as a parallel organization of outstanding engineers. It is autonomous in its administration and in the selection of its members, sharing with the National Academy of Sciences the responsibility for advising the federal government. The National Academy of Engineering also sponsors engineering programs aimed at meeting national needs, encourages education and research, and recognizes the superior achievements of engineers. Dr. Charles M. Vest is president of the National Academy of Engineering.

The **Institute of Medicine** was established in 1970 by the National Academy of Sciences to secure the services of eminent members of appropriate professions in the examination of policy matters pertaining to the health of the public. The Institute acts under the responsibility given to the National Academy of Sciences by its congressional charter to be an adviser to the federal government and, upon its own initiative, to identify issues of medical care, research, and education. Dr. Harvey V. Fineberg is president of the Institute of Medicine.

The **National Research Council** was organized by the National Academy of Sciences in 1916 to associate the broad community of science and technology with the Academy's purposes of furthering knowledge and advising the federal government. Functioning in accordance with general policies determined by the Academy, the Council has become the principal operating agency of both the National Academy of Sciences and the National Academy of Engineering in providing services to the government, the public, and the scientific and engineering communities. The Council is administered jointly by both Academies and the Institute of Medicine. Dr. Ralph J. Cicerone and Dr. Charles M. Vest are chair and vice chair, respectively, of the National Research Council.

www.national-academies.org

Committee on Comparative Innovation Policy: Best Practice for the 21st Century*

William J. Spencer, *Chair*
Chairman Emeritus, *retired*
SEMATECH

Kenneth Flamm, *Vice Chair*
Dean Rusk Chair in International
  Affairs
Lyndon B. Johnson School of
  Public Affairs
University of Texas at Austin and
  STEP Board

Mary L. Good, *Vice Chair*
Donaghey University Professor
Dean, Donaghey College of
  Information Science and Systems
  Engineering
University of Arkansas at Little Rock
  and STEP Board

Alice H. Amsden
Professor of Political Economy
Massachusetts Institute of
  Technology

Bronwyn Hall
Professor of Economics
University of California at Berkeley

Mark B. Myers
Visiting Professor of Management
The Wharton School of Business
University of Pennsylvania

Gail H. Cassell
Vice President, Scientific Affairs
Distinguished Lilly Research Scholar
  for Infectious Diseases
Eli Lilly and Company

Alan Wm. Wolff
Managing Partner
Dewey Ballantine

Lewis S. Edelheit
Senior Research and Technology
  Advisor, *retired*
General Electric

*As of December 2006.

Project Staff*

Charles W. Wessner
Study Director

Sujai J. Shivakumar
Senior Program Officer

McAlister T. Clabaugh
Program Associate

David E. Dierksheide
Program Officer

Paul Fowler
Senior Research Associate

Ken Jacobson
Consultant

Jeffrey C. McCullough
Program Associate

---

*As of December 2006.

For the National Research Council (NRC), this project was overseen by the Board on Science, Technology and Economic Policy (STEP), a standing board of the NRC established by the National Academies of Sciences and Engineering and the Institute of Medicine in 1991. The mandate of the STEP Board is to integrate understanding of scientific, technological, and economic elements in the formulation of national policies to promote the economic well-being of the United States. A distinctive characteristic of STEP's approach is its frequent interactions with public and private-sector decision makers. STEP bridges the disciplines of business management, engineering, economics, and the social sciences to bring diverse expertise to bear on pressing public policy questions. The members of the STEP Board* and the NRC staff are listed below:

---

*As of December 2006.

STEP Staff*

Stephen A. Merrill
Executive Director

Charles W. Wessner
Program Director

McAlister T. Clabaugh
Program Associate

Sujai J. Shivakumar
Senior Program Officer

David E. Dierksheide
Program Officer

Jeffrey C. McCullough
Program Associate

Paul Fowler
Senior Research Associate

Mahendra Shunmoogam
Program Associate

---

*As of December 2006.

Contents

Preface

Recognizing that a capacity to innovate and commercialize new high-technology products is increasingly a part of the international competition for economic leadership, governments around the world have taken active steps to strengthen their national innovation systems. These steps underscore the belief that the rising costs and risks associated with new potentially high-payoff technologies, and the growing global dispersal of technical expertise, require national R&D programs to support new and existing high-technology firms within their borders.

We define innovation as the transformation of an idea into a marketable product or service, a new or improved manufacturing or distribution process, or even a new method of providing a social service. This transformation involves an adaptive network of institutions that encompass a variety of informal and formal rules and procedures—a national innovation ecosystem—that shape how individuals and corporate entities create knowledge and collaborate to bring new products and services to market. If we define competitiveness as the ability to gain market share by adding value better than others in the globalized economic environment, the ability of these actors to collaborate successfully within a given innovation ecosystem gains significance. Recognizing this, policymakers around the world are supporting a variety of initiatives to reinforce their national innovation ecosystems as a way of improving their national competitiveness.

The proliferation of national initiatives to support innovation highlights the need for better understanding by U.S. policy makers of the objectives, structure, operation, funding levels, and trends characterizing some of the major programs around the world. These programs and associated policy measures are of great relevance to the United States both for their potential impact on U.S. competitiveness and for the lessons they may hold for U.S. programs.

With these objectives in mind, the National Research Council's Board on Science, Technology, and Economic Policy (STEP) has embarked on a study of selected foreign innovation programs in comparison with major U.S. programs. As such, the premise of this study is not to consider the possibility of a pure *laissez-faire* approach to fostering innovation, but rather to recognize the importance of targeted government promotional policies relative to innovation.[1] The analysis, carried out under the direction of an ad hoc committee, is to include a review of the goals, concept, structure, operation, funding levels, and evaluation of foreign programs designed to advance the innovation capacity of national economies and enhance their international competitiveness.[2]

In his welcoming remarks as the chair of this study, William Spencer stated that the purpose of the study's inaugural conference held on April 15, 2005, "was to try to gather the facts on how innovation and technology transfer were being funded in the various economic regions, and in particular on the roles of private and public funding." In particular, the conference focused on how universities, laboratories, and the private sector—both large companies and small—can link together in an effective system of national innovation. This volume provides a summary of this conference.

THE CONTEXT OF THIS REPORT

Since 1991 the STEP Board has undertaken a program of activities to improve policy makers' understanding of the interconnections among science, technology, and economic policy and their importance to the American economy and its international competitive position. The Board's interest in comparative innovation policies derive directly from its mandate.

This mandate has previously been reflected in STEP's widely cited volume, *U.S. Industry in 2000,* which assesses the determinants of competitive performance

---

[1]Government programs to promote promising technologies are a well-known and longstanding practice. See, for example, Vernon W. Ruttan, *Technology, Growth, and Development: An Induced Innovation Perspective.* Oxford, UK: Oxford University Press, 2000.

[2]Thus, while cognizant of the role of Defense Advanced Research Projects Agency (DARPA), and more broadly the Department of Defense, in the U.S. innovation system, the focus of the conference was on civilian technology programs that operate closer to market than does DARPA. In addition, as Alic and Branscomb et al. have described in *Beyond Spin-off*, the earlier military driven model of U.S. innovation is no longer as effective as it once was. DARPA funding of advanced technologies, particularly in Information Technology (IT), have had enormous impact, although largely on platform technologies that had wide and profound spillovers. Indeed the emergence of China and certainly India in the global economy attests to the impact of the Internet, to which DARPA made major contributions. See John A. Alic, Lewis M. Branscomb, Harvey Brooks, Ashton B. Carter, and Gerald L. Epstein, *Beyond Spin-off: Military and Commercial Technologies in a Changing World*, Boston, MA: Harvard Business School Press, 1992.

in a wide range of manufacturing and service industries, including those relating to information technology.[3] The Board also undertook a major study, chaired by Gordon Moore of Intel, on how government-industry partnerships can support the growth and commercialization of productivity enhancing technologies.[4] Reflecting a growing recognition of the importance of the surge in productivity since 1995, the Board also launched a multifaceted assessment, exploring the sources of growth, measurement challenges, and the policy framework required to sustain the New Economy.[5]

The current study on Comparative Innovation Policies builds on STEP's experience to develop an international comparative analysis focused on U.S. and foreign innovation programs. The analysis will include a review of the goals, concept, structure, operation, funding levels, and evaluation of foreign programs similar to major U.S. programs. Among other initiatives, this study will convene senior officials and academic analysts engaged in the operation and evaluation of these programs overseas to gain a first-hand understanding of the goals, challenges, and accomplishments of these programs.

The project held its opening event, "Innovation Policies for the 21st Century," on April 15, 2005. This international symposium drew experts from Europe, North America, and East Asia to provide overviews of major programs underway around the world to support innovation. This conference report summarizes their practical, "hands-on" insights concerning government and government-related programs that have worked. While the conference stimulated a rich and varied discussion, it did not (nor could it reasonably hope to) cover all facets of this important topic. For example, the issue of national treatment of intellectual property rights, while raised by some speakers, did not emerge as a focus of discussion. (The relationship between national innovation policies and global linkages is another issue touched on in this conference but not fully amplified. Similarly, the issue of national themes or innovation focus as practiced in different parts of the world was raised during the conference but not sufficiently articulated.) These issues are important and call for further attention. This report reflects both the strengths and limitations of the conference of April 15, 2005; it captures the scope and diversity of national programs and raises issues of direct policy interest for further research.

---

[3]National Research Council, *U.S. Industry in 2000: Studies in Competitive Performance,* David C. Mowery, ed., Washington, D.C.: National Academy Press, 1999.

[4]This summary of a multivolume study provides the Moore Committee's analysis of best practices among key U.S. public private partnerships. See National Research, *Government-Industry Partnerships for the Development of New Technologies: Summary Report,* Charles W. Wessner, ed., Washington, D.C.: The National Academies Press, 2003. For a list of U.S. partnership programs, see Christopher Coburn and Dan Berglund*, Partnerships: A Compendium of State and Federal Cooperative Programs,* Columbus, OH: Battelle Press, 1995.

[5]National Research Council, *Enhancing Productivity Growth in the Information Age: Measuring and Sustaining the New Economy,* Dale W. Jorgenson and Charles W. Wessner, eds., Washington, D.C.: The National Academies Press, 2007.

ACKNOWLEDGMENTS

We are grateful for the participation and the contributions of the Defense Advanced Research Projects Agency, the National Institute of Standards and Technology, the National Science Foundation, and Sandia National Laboratories.

We are indebted to Ken Jacobson for his preparation of this meeting summary. Several members of the STEP staff also deserve recognition for their contributions to the preparation of this report, including Sujai Shivakumar, McAlister Clabaugh and David Dierksheide for their role in preparing the conference and getting this report ready for publication.

NRC REVIEW

This report has been reviewed in draft form by individuals chosen for their diverse perspectives and technical expertise, in accordance with procedures approved by the National Academies Report Review Committee. The purpose of this independent review is to provide candid and critical comments that will assist the institution in making its published report as sound as possible and to ensure that the report meets institutional standards for quality and objectivity. The review comments and draft manuscript remain confidential to protect the integrity of the process.

We wish to thank the following individuals for their review of this report: Vinod Goel, The World Bank; Thomas Howell, Dewey Ballantine LLP; Kent Hughes, Woodrow Wilson International Center for Scholars; William Morin, Applied Materials; Dirk Pilat, Organization for Economic Cooperation and Development; and Andrew Toole, Rutgers University.

Although the reviewers listed above have provided many constructive comments and suggestions, they were not asked to endorse the content of the report, nor did they see the final draft before its release. Responsibility for the final content of this report rests entirely with the committee and the institution.

William J. Spencer   Charles W. Wessner

I

INTRODUCTION

Innovation Policies for the 21st Century

The capacity to innovate and commercialize new goods and services remains vital to the future competitiveness of the United States and indeed all participants in the global economy. Reinforcing and sustaining this capacity is particularly salient as research, development, manufacturing, and the delivery of services, made possible by new information and communications technologies, become ever more global. The emergence of new participants in the global economy, focused on attracting and developing high-technology industries within their national economies, is increasingly significant. China, for example, combines the advantages of high-skill and low-wage knowledge workers with a strong sense of national purpose. Responding to these structural changes in the global economy, other advanced economies have already initiated major programs, often with substantial funding, that are designed to attract, nurture, and support innovation and high-technology industries within their national economies. In this new competitive paradigm, the United States cannot assume that its continued preeminence in science and technology is assured.

As the National Academies noted in its recent report, *Rising Above the Gathering Storm*, "this nation must prepare with great urgency to preserve its strategic and economic security. Because other nations have, and probably will continue to have the competitive advantage of low-wage structure, the United States must compete by optimizing its knowledge-based resources, particularly in science and technology, and by sustaining the most fertile environment for new and revitalized industries and the well-paying jobs they bring."[1]

---

[1]National Academy of Sciences/National Academy of Engineering/Institute of Medicine, *Rising Above the Gathering Strom: Energizing and Employing America for a Brighter Future,* Washington, D.C.: The National Academies Press, 2007.

Responding to this challenge requires that we recognize that the nature and terms of economic competition are shifting as the United States cooperates and competes in a global economy.[2] U.S. policy makers need to be aware of the wide variety of innovation and competitiveness policies that many nations have adopted. These policies are designed to build research capacities and to acquire knowledge, and then to transition that knowledge directly to companies and support their development. The power of such well-financed and integrated national programs to shift the terms of international competition is often underestimated. In addition, other national programs are more modest in scale, providing essentially market-based incentives to encourage the transition of new technologies to the market. Yet, they too can have a significant impact on the terms of competition. A comparative perspective is necessary to help us understand what policies are succeeding and why, how selected policies might be successfully adapted in the U.S. context, and what existing U.S. programs might be enhanced.

Above all, it is important to understand, as one recent report notes, that the pace of competition is accelerating.[3] To better understand how competition is evolving, the National Academies' Board on Science, Technology and Economic Policy (STEP) held a symposium on April 15, 2005, which drew together leading academics, policy analysts, and senior policy makers from around the globe to describe their national innovation programs and policies, outline their objectives, and highlight their achievements.

This introductory essay summarizes the key issues raised at this National Academies symposium on Innovation Policies for the 21st Century. Contemporary approaches to innovation policy draw explicitly and implicitly on the idea of an innovation ecosystem, and Section A introduces this concept and the role of intermediating institutions in delivering the fruits of research to the marketplace. Section B highlights new competitive challenges related to the emergence of China and India as major new participants in the global economy. Section C looks at innovation programs and policies adopted by several developed nations to innovate and commercialize knowledge in today's global marketplace. Section D then reviews selected U.S. policies and programs designed to spur the commercialization of innovation. Finally, Section E draws together the need for a comparative perspective that draws on best practices in the United States and overseas.

---

[2]Kent Hughes has argued in this regard that the challenges of the 21st century require new strategies that take account of new technologies, new global competitors, as well as new national priorities concerning national security and the environment. See Kent Hughes, *Building the Next American Century: The Past and Future of American Economic Competitiveness,* Washington, D.C.: Woodrow Wilson Center Press, 2005, Chapter 14.

[3]Ibid.

"This symposium is about competitiveness: Some countries are tying to figure out how to get it, others how to keep it, and still others how to get it back. And it's all about learning how to move fast and win in a brutally competitive global economy, such as we've never seen."[a]

Dr. Lewis Edelheit
Senior Vice President of Corporate Research, General Electric, *retired*

[a]See comments by Lewis Edelheit in the proceedings section of this volume.

UNDERSTANDING THE INNOVATION ECOSYSTEM

How can we better capitalize on national investments in research? More specifically, how can we deliver the fruits of research through products and processes that both enhance welfare and generate wealth? And how do we generate the types of output from our universities and research centers that will help our regional economies grow and meet the challenges of the future? Beyond merely focusing on increasing inputs (such as more funds for basic research) on one hand or setting output targets and mandating results on the other, the innovation ecosystem approach examines the complex processes through which innovations emerge through a variety of collaborative activities to become commercially valuable products.[4]

Many of the speakers at the symposium drew on the idea of an innovation ecosystem. An *innovation ecosystem* is described below.

What Is an Innovation Ecosystem?

An innovation ecosystem captures the complex synergies among a variety of collective efforts involved in bringing innovation to market.[5] These efforts

---

[4]Drawing from presentations by the STEP Board staff to the PCAST, the concept of an innovation ecosystem was adopted by the President's Council of Advisors on Science and Technology and the Council on Competitiveness, among others. For an early articulation of the concept, see Charles W. Wessner, "Entrepreneurship and the Innovation Ecosystem," in David B. Audretsch, Heike Grimm, and Charles W. Wessner, *Local Heroes in the Global Village: Globalization and the New Entrepreneurship Policies*, New York: Springer, 2005.

[5]Consciously drawing on this ecosystems approach, the Council of Competitiveness' National Innovation Initiative (NII) report and recommendations address the need for new forms of collaboration, governance and measurement that enable U.S. workers to succeed in the global economy. Council on Competitiveness, *Innovate America: Thriving in a World of Challenge and Change*, Washington, D.C.: Council on Competitiveness, 2005.

include those organized within as well as collaboratively across large and small businesses, universities, and research institutes and laboratories, as well as venture capital firms and financial markets.[6] Innovation ecosystems themselves can vary in size, composition, and in their impact on other ecosystems. The strength of the linkages across a given innovation ecosystem can also vary.

Beyond this description, the term "innovation ecosystem" also captures an analytical approach that considers how public policies can improve innovation-led growth by strengthening links within the innovation ecosystem. Intermediating institutions (such as public-private partnerships) can play a key role in this regard by aligning the self-interest of venture capitalists, entrepreneurs and other participants within a complex innovation ecosystem with desired national objectives.[7]

The idea of an innovation ecosystem builds on the concept of a National Innovation System (NIS) popularized by Richard Nelson. According to Nelson, a NIS is "a set of institutions whose interactions determine the innovative performance . . . of national firms."[8] Too often, however, analysts and policy makers tend to see the innovation system as a static concept—a historical "given." To some extent, this is true. Innovation systems, at least initially, are normally not consciously developed for the purpose of enhanced competitiveness; rather they evolve from a vast array of loosely related institutions and policies. By contrast, the idea of an ecosystem evokes our understanding of complex and dynamic interdependencies in the natural world. In biology, ecosystems refer to interdependencies among particular plant and animal communities and the nonliving physical environment that supports them.[9] Taken together, the idea of a national

---

[6]In his luncheon address, John Marburger noted that a recent report by the President's Council of Advisors on Science and Technology identified five major categories of institutional participants. See President's Council of Advisors on Science and Technology, "Sustaining the Nation's Innovation Ecosystem," Washington, D.C.: Executive Office of the President, June 2004.

[7]National Research Council, *Government-Industry Partnerships for the Development of New Technologies: Summary Report*, Charles W. Wessner, ed., Washington, D.C.: The National Academies Press, 2003.

[8]See Richard R. Nelson and Nathan Rosenberg, "Technical Innovation and National Systems," in *National Innovation Systems: A Comparative Analysis*, Richard R. Nelson, ed., Oxford, UK: Oxford University Press, 1993. Nelson notes that the idea of a "national innovation system" captures "a new spirit of what might be called 'techno-nationalism' . . . combining a strong belief that the technological capabilities of national firms are a key source of competitive prowess, with a belief that these capabilities are in a sense national, and can be built by national action" (p. 5). The National Innovation System model appeals to policy makers since it provides an interpretive scheme that focuses on the nation as a unit of analysis. For a critique of the nation as a unit of analysis, see John de la Mothe and Gilles Paquet, "National Innovation Systems, 'Real Economies' and Instituted Processes," *Small Business Economics* 11:101–111.

[9]For an early definition of "ecosystem," as incorporating animal and plant systems in the context of other inorganic and living components, see A. G. Tansley, "British Ecology During the Past Quarter Century: The Plant Community and the Ecosystem," *The Journal of Ecology* 27(2):513–530. See also Henry Chesbrough, *Open Innovation: The New Imperative for Creating and Profiting From Technology*, Cambridge, MA: Harvard Business School Press, April 2003.

innovation "ecosystem" draws particular attention to the complex processes, interactions, and network relations taking place within a real economy.[10]

The idea of an innovation ecosystem thus highlights the multiple institutional variables that shape how research ideas can find their way to the marketplace. These include, most generally, rules that protect property (including intellectual property) and the regulations and incentives that structure capital, labor, and financial and consumer markets. A given innovation ecosystem is also shaped by shared social norms and value systems—especially those concerning attitudes towards failure, social mobility, and entrepreneurship.[11] (See Box A.) Innovation ecosystems are also conditioned by interest rate and exchange rate structures found within modern economic systems. Importantly, innovation ecosystems can also be strengthened by developing new institutional mechanisms that create new patterns of interaction, market knowledge, and incentives that motivate new entrepreneurship.

Fostering Local Innovation Ecosystems

A national innovation ecosystem is made up of a network of local innovation ecosystems. In an economy as large and complex as that of the United States, these local innovation ecosystems are themselves often significant. In his luncheon address to symposium participants, John Marburger, the Science Advisor to the President drew on his own experience in creating a university-based research park as president of the State University of New York at Stony Brook to summarize five principles that he viewed as necessary for fostering vibrant local innovation ecosystems.

1. Build competencies with attention to regional strengths. This consideration, he noted, is important for a large country like the United States, whose markets display very strong regional differences but each of whose regions possess their own strengths and possibilities. Institutions cooperating in regional development must hire people whose interests enhance and complement what is already found in the environment, which "doesn't happen unless somebody pays attention to it." The idea is to build *regional* strength, not just *institutional* strength. When several research institutions are located in the same region, they

---

[10]The emerging NIS literature draws attention to the presence of interactions and flows among public and private sector organizations in initiating, modifying, and diffusing new technologies. See P. Patel and K. Pavitt, "National Innovation Systems: Why They are Important and How They Might be Compared?" *Economic Change and Industrial Innovation*, 1994. See also C. Endquist, ed., *Systems of Innovation: Technologies, Institutions, and Organizations,* London, UK: Pinter, 1997.

[11]For a survey of attitudes towards entrepreneurship, see EOS Gallup Europe, *Entrepreneurship,* Flash Eurobarometer 146, January 2004. The survey shows, among other details, that Europeans have a greater fear of entrepreneurial failure—including loss of property and bankruptcy—than do Americans. Accessed at *<http://ec.europa.eu/enterprise/enterprise_policy/survey/eurobarometer146_en.pdf>*.

BOX A
National Attitudes and Support for
Innovation and National Industries

A significant transnational comparative issue that emerges from the conference concerns national attitudes or ideologies affecting innovation policies.

In the case of innovation programs, for example, Finland's Dr. Kotilainen characterizes government R&D funding, including payments to private industry, as "investment" rather than "expenditure." In contrast, Canada's Dr. Nicholson reports that the public discourse of innovation programs in his country emphasize "repayability" of government expenditures. As in the United States, critics in Canada often denounce government innovation programs as "corporate welfare," charging that they interfere with the market and are too focused on large companies.[a]

In the case of tax policies, China and Taiwan have created tax free (and even negative tax) environments for some high-technology sectors. As Thomas Howell points out in his analysis of China's semiconductor industry, the magnetic effect of such policies have been considerable.

In the United States, such treatment is often not politically feasible for profitable high-tech manufacturing, although state governments often make substantial tax concessions to attract and retain businesses.[b] From a U.S. perspective, the key point is that in many countries, the development of high-technology industry, with the growth in wages and jobs it entails, has the same broad political consensus that U.S. policy reserves for defense expenditure.

---

[a]While widespread, these views understate the role the government has often played in developing new high-technology industries. See Vernon Ruttan, *Technology, Growth and Development: An Induced Innovation Perspective*, Oxford, UK: Oxford University Press, 2002. Large companies, like small ones, often face a "Valley of Death" for new ideas. For a perspective from a large company (General Electric) on the challenges new technologies face in large companies, see National Research Council, *The Advanced Technology Program: Assessing Outcomes*, Charles W. Wessner, ed., Washington, D.C.: National Academy Press, 2001, p. 96; for a discussion of "picking winners and losers," see p. 51.

[b]By comparison, extractive industries such as mineral, oil, gas, and the agricultural industry often do benefit from favorable tax treatment and multiple direct and indirect subsidies in the United States and elsewhere. To some extent these differing measures reflect historical events and the effort to maintain farm incomes in commodity markets

---

benefit by cooperating in recruitment and group development. Stony Brook, Cold Spring Harbor Lab, and Brookhaven National Lab, for example, share information on an informal basis about areas of concentration and often collaborate on recruitment.

2. Identify a research strategy. Stony Brook's conscious decision to make biomedical research a priority meant allocating university resources to proposals and projects that work together to build a foundation for future successes—even if, "in terms of some sort of absolute measure of quality," these were at times not the best proposals to come forward. While there were exceptions to this practice, a bias was maintained in favor of those fields that could be expected to help further the overall strategy. "That requires leadership," Dr. Marburger declared. "It does not happen in a university environment unless someone is willing to push on it." Faculty development and capital improvements were coordinated to enhance biotechnology capabilities. While other areas needed and deserved attention, the immediate opportunities for funding lay in the biosciences, which therefore received the focus.

3. Build a regional environment. In the early 1980s, Long Island business organizations were not aware of the rapidly growing opportunities in the biotechnology industry. They did not appreciate the significance of an emerging major tertiary health-care facility or the value of federal funding as a source of technology. The Long Island economy was then dominated by large aerospace contractors—principally, Grumman Corporation—that was to fall by the wayside as the cold war came to an end and industry shifted completely. "So it was important for me and my counterparts at the two laboratories to get together, pound the pavement, and talk to people—to take the biotechnology message to business groups, chambers of commerce, and state and local government agencies," Dr. Marburger recalled. "The whole region had to cooperate in making this work, and somebody always has to take the first step to get others together." Because Long Island's business community was aware of the dangers of relying on a single industry, these efforts by the leading centers of research to work together with business were warmly received.

4. Form regional partnerships. Institutional rivalries are counterproductive; cooperation and collaboration are essential for regional-scale development; and regional-scale development is important for a stable pattern of growth. The fact that companies start up, grow, then frequently either die or move elsewhere is not necessarily the end of the world, but it does necessitate continual start-ups. Some of the new companies may survive and add permanently to the economy, some may have to be replaced with others that are sufficiently similar to stabilize the workforce. It is because regional partnerships enhance mobility and multiply opportunities for workers and for businesses that a critical mass of mutually compatible businesses is needed to stabilize the inevitable effect of startups' moving away. "In Silicon Valley in its heyday, and it is presumably still somewhat like this, you had the phenomenon of frequent moves of technical personnel from one company to another," Dr. Marburger observed. "There was a great deal of mobility—companies came and went, started and failed—and in general the makeup of the workforce was similar, which stabilized employment in the area despite the dynamics in the companies."

5. Fund the machinery, which consists of facilities, people, and organizations. None of this happens without people who know that their job is to make it happen; neither regional development nor technology transfer can be made to work with volunteers. "I travel around the country looking for regions that are succeeding," Marburger said, "and many are attempting to do it on a voluntary basis, but only those where there is some sort of executive center with a paid workforce [are having success]." In other words, whether at a state, county, or local government economic development office, or at an organization that is either freestanding or associated with a university or a business group, someone has to know that technology transfer is his or her job. Technology-related economic development usually entails investing state and local government funds in facilities so as to reduce costs for startup tenants, and people are needed to bring entrepreneurs together with financial and technical support. Nearly always, such people are more than brokers. They are teachers and counselors, too: for entrepreneurs, who know the technology but not business practices, and for investors, who are ignorant of the ways of engineers and scientists.

Concluding his address, Dr. Marburger acknowledged that while these lessons may not apply to every situation, the support for university-based research parks, and of research parks based around a nucleating asset other than a university, is growing, thriving, and becoming an important part of the U.S. innovation ecology.

Complementing Marburger's perspective on developing successful local innovation ecosystems, other participants described policy efforts at the national level to develop national innovation potential and competitiveness. As participants learned at the conference, these efforts are being undertaken by reemerging powers such as China as well as established U.S. allies and competitors like Germany and Japan. We look next to how selected speakers at the symposium characterized these challenges.

THE RISE OF NEW COMPETITORS

In his opening address, Carl Dahlman of Georgetown University noted that, while the United States is the world's preeminent economy, accounting for more than a quarter of the world's gross domestic product today, "other nations are catching up fast." Several developing countries in Asia are investing heavily in education and are building world-class science and technology infrastructures. Some nations are also acting decisively to attract and retain important high-technology industries within their borders, as seen in China.

The Challenge from China: National Policy with Purpose

In particular, Dahlman noted that the technology level and the scale of Chinese industry continue to grow at a very high rate, challenging strategic calculations of competitors around the world.[12] China's economy, he observed, grew at about 8 to 10 percent per year over the previous two decades.[13] Its "gigantic" internal markets afford it a very important strategic advantage in negotiating externally, as evidenced by the fact that foreign investors have been willing "to bring in not the second or third rate technology, but the very best" for application to their operations in China.

China's competitive advantages, according to Dahlman, include:

- A very high savings and investment rate (about 40 percent) compared with the rest of the world (about 20-plus percent);
- Skill in tapping into global knowledge both through direct foreign investment and the Chinese Diaspora;
- A critical mass in R&D that is increasingly deployed in a very focused effort to increase its competitiveness;[14]
- A large and growing manufacturing base combined with advanced export-trade logistics;
- Continuing strong investments in education and training, endowing China with the world's third-largest scientific and technical work force focusing on R&D;
- A very large supply of excess labor in the agricultural sector (some 150 million-200 million people) that continue to keep down labor costs;
- A government with a very strong sense of national purpose, which "helps to coordinate things, although it creates some other kinds of problems."

China, noted Carl Dahlman, has demonstrated the "importance of the nation-state" not only in implementing development plans and visions but also in providing a stable macroeconomic framework. He underlined what he called the "tremendous pragmatism" exhibited by the Chinese government in setting up needed

---

[12]See Carl Dahlman and Jean-Eric Aubert, *China and the Knowledge Economy: Seizing the 21st Century*, Washington, D.C.: The World Bank, 2001.

[13]This trend is expected to continue in the near future. See Reuters, "China Sees No Quick End to Economic Boom," February 21, 2006.

[14]On February 9, 2006, China's cabinet listed 16 key technologies to receive more support from government and private industry. These included computer software, telecommunications, nuclear energy and a military-managed space program. To speed progress in these areas, the cabinet announced that research and development spending should rise dramatically to reach 2.5 percent of gross domestic product by 2010. In 2004, R&D spending was 1.23 percent of GDP, according to a Chinese ministry official from the statistics department. *The Washington Post*, "Chinese to Develop Sciences, Technology," February 10, 2006, p. A16.

incentives within the Chinese innovation system: "Although it is supposed to be a communist system, they have stock incentive plans in the research institutes." Similarly remarkable, he noted, is that one-third to one-half of the cost of higher education is paid by the students through tuition. While the Chinese have been focusing on technology and education for the previous two decades, the policies currently in development were more coordinated than those preceding them. "They are just really revving this up even more," he commented.

China Grows a Semiconductor Industry

In his presentation, Thomas Howell illustrated how the focus and coordination of China's innovation policies has resulted in the rapid development of a world-class semiconductor industry. Howell noted that China's rise in semiconductors is all the more significant when one considers that much of that nation's science infrastructure was destroyed during the decade of the Cultural Revolution (1966-1976). The previous command economy model ensured that Chinese technology remained 10 to 15 years behind the global state of the art, he added.

China introduced market reforms to its command economy in the early 1980s. However, around the beginning of its tenth Five-Year Plan in 2001 and concurrent with joining the World Trade Organization (WTO), China fundamentally reappraised its command economy and "essentially decided to jettison the whole system." While retaining the economic nationalism that suffused all its earlier Five-Year Plans, China has largely abandoned the command method in favor of a system that uses incentives permitted under the WTO, including subsidies, tax measures, targeted government procurement, and the like.

Simultaneously taking place is a thorough decentralization, with most of the policies being implemented locally rather than at the national level; a fundamental redefinition of the industry-government relationship, with an emphasis on the independence of enterprises' decision making; and liberalization of inward investment permitting foreign companies to establish fully owned subsidiaries. Tariffs were eliminated. And pressure to transfer technology eased, although that pressure has not ceased entirely.

These measures added up to a "paradigm shift" in which Chinese planners abandoned their own command system and embraced Taiwan's state-directed market, said Howell. He displayed a table showing that virtually every current Chinese policy in the semiconductor field has a Taiwanese antecedent;[15] indeed, many of them were implemented with the assistance of Taiwanese advisers.[16] The

---

[15]Another duplication of Taiwanese policy was to allow companies to exist in a tax-free environment.

[16]For example, one of the leaders in setting up Hsinchu Park, Irving Ho, acted as a consultant on the industrial parks that have been built on the mainland in the previous 5 years.

Policy/Practice	China 1994: Command Economy Mode	Taiwan 2000: Partnership Model	China 2002
Principal form of leading semiconductor enterprises	State-owned enterprise	Private, government holds passive minority share	Private, government holds passive minority share
Business model of leading semiconductor firms	Integrated device maker	Foundry	Foundry
Policy toward foreign direct investment	Heavily restricted	Liberalized	Liberalized
Promotion of IC design industry	Emphasis on state-owned research institutes	Privatization of government research institutes Financial assistance to private companies	Privatization of government research institutes Financial assistance to private companies
Government as direct investor in leading firms	100% government ownership of semiconductor enterprises	Government passive minority equity stake	Government passive minority equity stake
Tariffs on semiconductors	6-30 %	0	0
Industrial parks	Over 100 "high-tech parks"	1 flagship park (Hsinchu), 2-3 others emerging (Tainan, Nankang)	1 flagship park (Zhangjiang), 2-3 others emerging (Suzhou, Beijing)
Major financial incentives to individuals	None	Major tax benefits	Major tax benefits
Government controls enterprise decision making	Yes	No	No
Government promotion of venture capital sector	No	Yes	Yes

FIGURE 1 Microelectronics: China embraces Taiwan's model.
SOURCE: Presentation by Thomas R. Howell, Dewey Ballantine, "New Paradigms for Partnerships: China Grows a Semiconductor Industry," in Panel III of this volume.

function of China's central government in policy became "mostly hortatory," he added, with the actual benefits and promotional measures implemented largely by the regional governments and local governments in line with the central government's intentions (see Figure 1).

According to Howell, the Chinese also acted to effectively leverage their large internal market.[17] Notably, China emphasized its market's pull by applying in 2000 a differential value-added tax (VAT) that gave semiconductor devices manufactured in domestic fabrication plants a 14 percent cost advantage over imports into the Chinese market. Taiwanese companies, seeing that they might be shut out of China's growing market unless they invested there, rushed across the Strait. Although the VAT measure was subsequently withdrawn, that investment that rushed over remained.[18]

[17]By 2004, China's integrated circuit consumption reached $35 billion or nearly 20 percent of the worldwide demand. This was an increase of $10 billion from 2003. The corresponding 2004 China integrated circuit industry revenues were $6.6 billion, up $2.4 billion from 2003. See PricewaterhouseCoopers, "China's Impact on the Semiconductor Industry: 2005 Update, " 2006.
[18]*Federal Register Notice*, "2004 WTO Dispute Settlement Proceeding Regarding China: Value-Added Tax on Integrated Circuits," Wednesday, April 21, 2004.

As a result of these policies, the Chinese semiconductor industry has expanded faster than in any of the world's major economies. Howell noted that it grew at a rate of 40 percent in 2004 and is expected to achieve a compound annual growth rate of over 20 percent for the period 2002-2008, compared to 7.3 percent for the United States and 13.8 percent for Taiwan. According to Howell, China's semiconductor industry was valued in 2005 at about $24 billion[19] and is expected to grow to something on the order of $65 billion by 2007.[20]

As the bulk of wafer-fab investment moves to China—and Howell projected that China will boast some 30 new fabrication plants in the ensuing 3 years compared to 6 in the United States—China is likely to attract more science and engineering graduates from around the world (many of Chinese descent) and develop into the world's premier locus of semiconductor design and manufacturing. Given that semiconductors are the enabling technology of the modern information and communication age, this poses a major competitive as well as strategic challenge to the United States, he concluded.

Some Challenges Facing China

Although some believe the Chinese juggernaut to be unstoppable, Carl Dahlman outlined four key internal challenges to China's continuing growth performance.

- **China's Closed Political System.** Although China is moving more and more toward a market economy, it does not have a democratic political system. "At some point there is tension between people's willingness to live in a more constrained system as opposed to a freer one," Dahlman observed, saying it is not easy to predict how this issue will play out.
- **Growing Economic Inequality.** Inequality is growing in China among both people and regions, and it is becoming a serious concern.
- **A Vulnerable Financial System.** China's many nonperforming loans may not be a problem if the economy continues to grow very fast, "but if it slows down, then the relative size of the non-performing loans is a big problem."
- **Natural Resource Constraints.** On a per capita basis, China's natural resources were quite thin. China is a highly energy-dependent nation, a problem it has been addressing by using its large foreign currency reserves to acquire access to raw materials around the world.

---

[19]Throughout this volume all dollars are U.S. unless otherwise indicated.

[20]China produced approximately 30 billion integrated circuit (IC) chips in 2005, a year-on-year increase of 36.7 percent. The sector recorded a sales volume of 75 billion yuan (US$9.2 billion), up 37.5 percent over the previous year, Xinhua News Agency, January 28, 2006.

India's Growing Potential

India has seen its annual growth rate rise from the 2 to 3 percent that was traditional prior to the past decade through the 5 to 6 percent level to around 8 percent. It is, in Carl Dahlman's words, "poised to do a China," held back only by its own internal constraints.[21] Chief among these, he noted, is a surfeit of bureaucracy stifling a flair for entrepreneurship. Political wrangling over the rural-urban divide in this democracy has also stalled the development of much needed transportation facilities and other urban infrastructures needed to capitalize on current opportunities for growth.[22] Still, India possesses major advantages, including a vibrant entrepreneurial class and a critical mass of capable, highly trained scientists and engineers, most notably in the chemical and software fields.[23] The Indian Diaspora also maintains linkages back to the home market from overseas. These advantages, as well as the presence of a large English-speaking population, have already made India a major locus for outsourcing of business processes as well as an attractive place for multinational corporations to conduct R&D.

In fact, because of India's tremendous cost advantage in human capital, foreign firms are increasingly locating large R&D facilities in India.[24] In addition, Indian companies such as Wipro were beginning to do contract research in India on behalf of multinationals in pharmaceuticals as well as in information and communications technology. Dahlman added that India has relatively deep financial markets compared to other developing countries, and, under the pressure of China's liberalization, is finally beginning to look not just internally but also outside. It is also seeking strategic alliances, aided by success in capitalizing on its own diasporas for access to information and markets.

Dahlman noted that one of the main lessons to be drawn from the Indian experience is the significance of the long term: The investments in high-level human capital that were now beginning to pay off for India were made as far back as Prime Minister Nehru's time in the 1950s through the mid-1960s. The Indian Institutes of Technology and Indian Institutes of Management, world-class institutions that accepted only about 2 percent of applicants, have helped build

---

[21]Carl Dahlman and Anuja Utz, *India and the Knowledge Economy: Leveraging Strengths and Opportunities*, Washington, D.C.: The World Bank, 2005.

[22]*The Financial Times*, "India Needs Big Infrastructure Drive," February 23, 2006.

[23]For an analysis of India's economic potential compared to China, see Yasheng Huang and Tarun Khanna, "Can India Overtake China?" *Foreign Policy*, July–August, 2003. The authors argue that India's development strategy, while initiated later than China's and thus lagging China, is more sustainable because it is more strongly based on fostering bottom-up entrepreneurial capacity.

[24]For example, IBM has recently announced its $6 billion investment in R&D in India. Saritha Rai, "India Becoming a Crucial Cog in the Machine at I.B.M." *The New York Times*, June 5, 2006. Other U.S. companies recently making large investments in Indian R&D capabilities include Microsoft, Qualcomm, and SAP.

a "truly gigantic pool" of world-class talent.[25] Galvanizing this skill pool could yield major competitive strengths. Harnessing the Indian Diaspora, so that the brain drain could be turned into a "brain gain," could also play a major role in India's development.

Key reforms needed for India to sustain its growth momentum, concluded Dahlman, include moving away from a very autarchic system to become a more integrated part of the global system, which will offer significant benefits from specialization and exchange and further reforming the legal and regulatory regimes, which continue to act as a brake on India's growth.

NEW INNOVATION POLICY MODELS
AMONG ESTABLISHED ECONOMIES

What are the implications for the United States? For the United States to remain competitive in the emerging competitive landscape, perhaps the main lesson is that it must pay attention. Important policy experiments are now underway in the advanced economies of Europe, Canada, and East Asia, and collectively they are shaping the conditions of international competition. These countries face challenges in innovation policy similar to those faced by the United States and, in some cases, share cultural attributes that might make elements of their innovation policies adaptable in the United States. While these models are not necessarily replicable in the American context, their descriptions at the conference did demonstrate the sustained, high-level policy attention that innovation policy receives abroad. Many of the programs have common objectives, and in some cases, common features. To capture the main features of these programs and their objectives, experts from Finland, Germany, Canada, Japan and Taiwan described their innovation programs and national policy initiatives. These are summarized below.

Finland's Successful Innovation Model

Finland is an example of a small country whose commitment to innovation policy and R&D investment has enabled it to become a global leader in high technology. Finland's Heikki Kotilainen[26] began by identifying key structural challenges facing his country, including globalization and the movement of manufacturing to Asia, serious demographic changes, and the need for environmental sustainability in an economy that was traditionally reliant on forestry products. Finland has faced the need to be innovative to overcome these challenges, he noted, adding that responding to the rapidly changing dynamics of innovation in itself constituted a challenge.

---

[25]Kanta Murali, "The IIT Story: Issues and Concerns." *Frontline*, 20(3), February 1, 2003.
[26]Then, the Deputy Director-General of Tekes, the highly regarded Finnish Technology Agency.

During the post-cold war economic crisis of 1992, Finland made a collective national decision, partly as a result of real investments, investments in education, research, and new technologies. Finland has successfully transformed its industrial structure from one based predominantly on natural resources to a more diversified portfolio that includes significant investments in the electronics and telecommunications sectors. Today, Finland—a country of 5.2 million people—is ranked by a variety of measures as second only to the United States in science and technology.[27] According to Heikki Kotilainen, this feat has been made possible through "conscious and continuous" investment and through the evolution of Finnish policy in the realm of science and technology.

Finland has increased its R&D spending as a percentage of GDP from 1.5 percent in 1985 to nearly 3.5 percent at the beginning of the current decade. While private sector spending has shown the most growth over this period, accounting for some 70 percent of current total investment, the public sector has been a prime mover. A key observation: Only when the government began to increase investments in R&D and related institutions did private investment follow.

This seriousness of purpose is reflected in Finland's public organizations in the R&D domain. The Academy of Finland is charged with funding basic research while Tekes (Finland's technology agency) is charged with funding applied research. Public sector R&D actors also include universities; VTT, a large multidisciplinary research institute; and a high level government council. The latter—called the Science and Technology Council—is a key element of the Finnish innovation policy system. This Council is chaired by the prime minister and includes key ministers; the directors-general of the Academy, Tekes, and VTT; as well as representatives of universities, industry, and unions. Together, they set out policy recommendations, revised every 3 years, based on reevaluations of Finland's strategic challenges and opportunities. Based on this outline, lending organizations such as Tekes cooperate with its industry and university partners to develop operational plans.

Reflecting national perceptions of the need for innovation and its effectiveness, Tekes itself has enjoyed a steadily rising budget, reaching approximately 430 million euros in 2005. Research funding in the form of grants and company funding in the form of both grants and loans are distributed through a variety of instruments. While he acknowledged these instruments themselves are not unique to Finland, Kotilainen added that Tekes' strength lay in its emphasis on implementation. This results-oriented approach places importance on cooperative networks between companies and universities so as to integrate technology

---

[27]Dr. Kotilainen notes that this claim is based on 2003 statistics published by the World Economic Forum and the IMD World Competitiveness Center and well as the 2001 UNDP Human Development Report. In turn, these statistics take into account measures such as per capita R&D levels, the quality of university and K–12 education, wireless and broadband penetration, and the presence of institutions such as SITRA and Tekes that facilitate cooperation among academia, industry, and government.

transfer into the process. Industry leadership and cost sharing, he added, are key elements to Tekes' success. A major cultural advantage is the high degree of trust, and the resulting low administrative overhead, required to make selections and process funding.

The impact of Finland's innovation institutions has been impressive, generating a remarkably sharp increase in Finland's high-technology exports—from below 5 percent in 1988 to over 20 percent in 1998. Reflecting the results-oriented perspective, Kotilainen reported that coinciding with Tekes' investment of 409 million euros in 2004, 770 new products reached the market and 190 manufacturing processes were introduced and that 720 patent applications, 2,500 publications, and 1,000 academic degrees were funded—reflecting, perhaps, that some Tekes awards may well be closer to the market than others. In addition, he noted that the receipt of Tekes funding has often caused project goals to be reset higher, and has caused project implementation to be speeded up in many cases. Finally, he noted that Tekes plays a major role in helping entrepreneurs surmount risk barriers: A study by Finland's National Audit Office in 2000 found that 57 percent of projects would not have been undertaken without the support provided by Tekes.

Germany: New Innovation Policies in a Federal Context

Despite its low growth rates, Germany remains an economic powerhouse, and a leading world exporter. At the same time, there are emerging vulnerabilities, including a high degree of dependence on its automotive cluster and an anticipated shortage in the supply of highly qualified labor. Still, Stefan Kuhlmann of the Fraunhofer Institute argued that Germany continues to be highly "innovation oriented." Germany's gross R&D expenditures are about 55 billion euros, or around 2.5 percent of GDP, with companies accounting for two-thirds of this expenditure. He added that Germany's 14.9 percent of the world market for R&D intensive goods placed it second to the United States and that it is in the European Union's top three in share of manufacturing sales attributed to new products. With 127 patent applications per inhabitant, Germany is the second highest among large countries, and ranked third among all nations in international publications, with 9 percent of the total.

Germany's innovation system is complex, with major decision making at the federal, Länder, and regional levels, involving much overlap of programmatic responsibilities. At the national level, both the Federal Ministry of Economics and Labor (BMWA) and the Federal Ministry of Research and Education (BMBF) fund a broad variety of technology and innovation programs—so broad, observed Stephan Kuhlman, as to be difficult to track sometimes. The Länder are responsible for funding and operating the nation's universities, but are increasingly going beyond this traditional role to set up programs designed to spur cooperative R&D, encourage partnering, develop incubators and science and technology parks, and furnish venture capital and loan guarantees. An additional source of R&D initiative and funding are the European Union programs although they are not as significant for Germany as for the smaller nations in Europe. Nonetheless, Kuhlmann observed that EU funding has contributed to R&D in the information and communications sector in Germany.

To take two concrete examples, Kuhlman reviewed BMWA's Pro Inno and BMBF's Inno Regio programs in more detail. Pro Inno has been in operation for more than 10 years and has invested 630 million euros between 1999 and 2003 with the goal of increasing R&D capability and SME (small- and medium-sized enterprises) competence. Subsidies under Pro Inno range between 25 and 50 percent of the cost of R&D personnel ranging across four program lines—cooperation with firms, cooperation with research organizations, R&D contracts and personnel exchange—with multiple applications totaling up to 350,000 euros per firm allowed. Since 1999, 4,870 firms and 240 research organizations have participated with 4,000 R&D employees per year engaged in Pro Inno projects. A 2002 evaluation showed that nearly three-fourths of participating firms would not have conducted R&D in the absence of this program.

The Inno Regio program is designed to strengthen the endogenous innovation potential of weak regions in eastern Germany by setting up sustainable innovation networks. The program encompasses not only SMEs, large companies, and research organizations, but also may other public and private activities and initiatives, funding both network management projects and projects aimed at developing products and services. The program is run as a three-stage competition—a qualification round (444 initiatives selected in 1999), a development round 25 out of 444), and a realization round where winners (23 of 25) receive multiyear financial support for their initiatives. An increase in innovation activities was observed under Inno Regio—two-fifths of the firms selected received patents and almost all introduced new products and, since 2000, 50 new firms have been founded. In Kuhlman's judgment, the program's main success is the creation of innovation networks across eastern Germany that brings together both public and private actors.

Finally, Kuhlman noted that Germany is trying to introduce more coordination and collaboration across agencies responsible for innovation policies. A "Partnership for Innovation" has been launched recently with the aim of improving the framework for innovation through the collaboration of public and private actors. A key initiative under this program has been a "High-tech Master Plan" to ease access to venture capital through the launch in early 2005 of a 10 million euro fund for start-ups.

Overall, Germany continues to launch new initiatives at the national level, but still suffers from limited resources per program, limited access to early stage finance and, above all, structural obstacles such as labor regulations that complicate the efforts of German entrepreneurs.

Canada: Strengthening Incentives to Attract Research Talent

Canada is a particularly interesting case in that it shares many social values with the United States. Many sectors of the Canadian economy are also highly integrated with those of the United States. At the same time, Canada has launched a wide variety of institutional and funding initiatives to encourage greater innovation, with positive results. Peter Nicholson of the Office of the Prime Minister noted that Canada's investments in a strong basic research capability are now paying dividends. Canada, with a population of roughly 32 million, spends around 19 billion U.S. dollars annually on R&D. Business expense on R&D account for about 55 percent of the country's total, government intramural expenditure on R&D is around 12 percent, with the remainder 33 percent of R&D spending coming from higher education. Given that Canada's innovation system is highly integrated with that of the United States, Canada's innovation programs have focused on building domestic capacity, largely by creating incentives to retain and attract scientific and research talent. These policies, described below, are now bearing fruit.

Traditionally, Canada's economy has been resource based, with much of its technical dynamism arising from its unique level of integration with the United States. "If we wanted to have something that was home-grown and that could give a degree of independence," he explained, "we [would have] had to build our innovation capacity from the ground up." The effort to build this foundation has been gained strength thanks to Canada's successful fiscal consolidation. With the federal budget in surplus since 1997, a "paradigm shift" in the federal government's support for higher education has taken place.

At the federal level, there are now four major innovation programs in place: (1) the Canada Research Chairs (CRCs), (2) the Canada Foundation for Innovation, (3) Technology Partnerships Canada, and the (4) Industrial Research Assistance Program (IRAP). With the exception of IRAP, these programs all started life in the 1990s.

Research Chairs. The objectives of the *Canada Research Chairs (CRCs)* were to attract and develop world-class researchers. The CRCs, which are awarded to a variety of disciplines (from Agriculture Engineering to the Visual Arts) are divided into two tiers.[28] The first is reserved for world leaders in their disciplines and provides an award of 7 year's duration, renewable indefinitely at $170,000 per year. These awards serve to sustain and attract world-class researchers to Canada. The second tier is to support "exceptional young faculty." It is renewable once and provides for $85,000 per year for 5 years. The program is successful,

---

[28]Additional information about the Canada Research Chairs can be accessed at <*http://www.chairs.gc.ca/web/about/index_e.asp*>.

Nicholson noted, moving from retaining faculty to recruiting well-qualified nominees from outside Canada. Over 1600 CRCs have been filled to date.[29]

Infrastructure. The *Canada Foundation for Innovation* is designed to set up and cofund leading-edge research infrastructure in universities and hospitals. The foundation has been endowed with a $3.1 billion grant from the federal government and has committed $2.5 billion in funding for 4,000 projects through competition based awards. The program's yearly budget is approximately $250 million. Like the CRCs, the Foundation's purpose is to attract and retain world-class researchers, promote collaboration and cross-disciplinary research, and foster strategic research planning, with the objective of transforming research and technology development in Canada.

Risk Share. The purpose of *Technology Partnerships Canada* is to risk-share industrial research and precompetitive development across a wide spectrum. Designed to address what Nicholson called a "persistent and frustrating" gap in Canadian firms' development of new technology, it covered from 25 to 30 percent of the costs involved in R&D, development of prototypes, and testing. In addition to significant cofinancing by industry, it featured cost recovery based on results. Targeting firms of all sizes, Technology Partnerships Canada focused its activities in aerospace and defense, environmental technologies, and enabling technologies including biotechnology and materials engineering.

Advising Services. The final technology program that Nicholson introduced is the *Industrial Research Assistance Program*, or IRAP. Funded at US$135 million a year, IRAP provides a range of both technical and business oriented advisory services and in some cases financial support to growth-oriented Canadian small and medium-sized enterprises. The program is delivered by an extensive integrated network of 260 professionals in 100 communities across the country. Working directly with these clients, NRC-IRAP is designed to support innovative research and development and commercialization of new products and services. The program is competitive, with only 20 to 25 percent of 12,000 applicants, receiving funding—with the average award at $30,000 per year with a maximum of $425,000 a year. Even so, Nicholson stated, the 3,000 or so funded projects have encouraged cooperation among subcontractors, suppliers, consultants, universities, and the Canadian National Research Council. Together, these programs

---

[29]The potential number of chairs allocated to each university depends on the proportion of research grants it wins in other national competitions, although a bonus is reserved for smaller institutions. Under the selection process, universities are expected to nominate candidates for the chairs in line with the same institution-wide strategic plan to which applications to Canada Foundation for Innovation (described below) must conform. Winners are selected by a three-person review panel or, in the absence of a consensus among the panel members, by a standing adjudication committee.

represent a comprehensive, yet market-oriented approach to strengthening related elements of the Canadian innovation system.

Japan: Restructuring for Resurgence

Despite its difficulties in the decade of the 1990s, Japan remains one of the world's premier technology powerhouses. As David Kahaner pointed out, Japan faces severe challenges, including a languid economy, an aging population, and stiff low-wage competition from China and other East Asian nations. Like Finland, Japan recognized the importance of institutional reform, leading to a variety of reforms aimed at reviving Japan's technological and commercial leadership. A major initiative in this respect, noted by Kahaner and Hideo Shindo of Japan's New Energy and Industrial Technology Development Organization (NEDO), is the S&T Basic Law of 1995, which provides a framework to improve economic development, social welfare, and environmental sustainability.[30]

A major outgrowth of this legislation is the founding in 2001 of Japan's Council for Science and Technology Policy (CSTP). The council is chaired by the Prime Minister and includes six cabinet ministers, five academics, and two industry representatives. The council is charged with developing a "grand design" for Japanese S&T policy. One of the CSTP's most important duties is drafting the country's 5-year S&T Basic Plan, which sets guidelines for the comprehensive and systematic implementation of Japan's overall S&T promotion policy. The goal of the first Basic Plan, which went into effect in 1996 and thus predated CSTP's creation, was to double government spending on R&D. The second Basic Plan, whose budget was set at $212 billion, is part of an effort to double the amount available for competitive funding.

Also in 2001, Japanese ministries were reorganized to streamline R&D funding and policy support. The administrative reforms include:

- The former Education Ministry and Science & Technology Agency were merged into the new Ministry of Education, Science, Culture, Sports, and Science & Technology (MEXT).
- A Ministry of Public Management, Home Affairs, Posts and Telecommunications—whose name was changed in 2004 to Ministry of Internal Affairs and Communications (MIC)—has arisen from the combination of the previous Management & Coordination Agency, Home Affairs Ministry, and Ministry of Posts & Telecommunications.
- The Ministry of International Trade & Industry (MITI) has been reborn as the Ministry of Economy, Trade & Industry (METI).

---

[30]Access Japan's S&T Basic Law of 1995 at <*http://www.mext.go.jp/english/kagaku/scienc04. htm*>.

Mr. Shindo also noted that Japan has in addition adopted new industrial policies to complement the organizational reforms set out in the Basic Science and Technology Plan. He cited the Nakagawa Report, published in 2004, which identified steps needed to establish and accelerate "the virtuous cycle of demand and innovation in order to bring about Japan's economic recovery and to create its future industrial structure." To this end, the Nakagawa Report draws together what Mr. Shindo described as a "very comprehensive" list of concrete policy priorities, including the identification of promising industrial areas (such as fuel cells and digital consumer electronics) and the development of policies for regional revitalization. The Nakagawa Report also considers cross-cutting policy issues such as the development of human resources (including continuing education for mature workers), intellectual property rights, research and development, standardization, and policies to encourage the development of new businesses by small- and medium-sized enterprises. Within a year of publication of the Nakagawa Report, Mr. Sindo said, NEDO with METI and the National Institute for Industrial Science and Technology (AIST) has developed a Technology Strategy Map to implement these plans.

Reflecting this increased commitment, Japan's total 2005 S&T budget rose to $36 billion, reflecting an increase of 2.6 percent over the previous year. Of that about $13 billion was for research expenses including researcher salaries, with the remainder financing infrastructure. Japan's R&D budget focuses on four key areas—nanotechnology and materials, information technology, life sciences, and the environment, with aerospace technology to be added as a fifth area. These investments are significant, reflecting Japanese goals of creating one thousand biotech companies and developing leadership in nanotechnology. Japan spends almost as much as the United States on an absolute basis on nanotechnology. Other areas of research focus include fuel cells, robotics, and computing research. (See Figure 2.)

Universities are a key element in current Japanese policy reforms. According to Kahaner, Japanese policy is seeking a larger role for university research, including collaboration among industry, academia, and government. In a major step designed to foster more technology transfer from universities, national universities have been converted over the past few years into independent administrative agencies. While still funded by the government, these agencies now have more autonomy and flexibility. For example, universities can now seek private funds and cooperate with industry. Laws have been enacted that allow Japanese professors to "become millionaires if they're good enough and have good enough ideas."[31] Over the past 5 years, many Japanese universities have established technology transfer offices, with technology management receiving unprecedented attention at Japanese universities. Japanese policy is also encouraging the forma-

---

[31]See presentation by David Kahaner in the proceedings section of this volume.

Program, bold font means competitive research funding	Ministry	2005 Budget in US$ Bil.
Life Science: **Molecular Imaging**	MEXT	1.150
Life Science: Research on Aids, Hepatitis, Emerging and Reemergi ng Infectious Disease	MHLW	4.526
Life Science: Efficient Breed Improvement Technology Based on Ge nome Breeding	MAFF	1.580
Information and Communication: Next Generation Back-bone	MIC	2.000
Information and Communication: **R&D for the Establishment of IT Infrastructure**	MEXT	2.974
Information and Communication: Human-assisting Robotics Realization Project	METI	0.900
Information and Communication: Autonomicus Movement Assisting Project	MLIT	0.490
Environment: **Promotion of Establishing Earth Observation System**	MEXT	1.017
Environment: Research on Agricultural, Forestal and Fishery Bio-cycle	MAFF	1.400
Environment: Development of Environmental Technology Based on Na notechnology	MEXT	0.400
Nanotechnology: **Fused Emerging Field Based on Nanotechnology and Materials**	MEXT	1.450
Nanotechnology: Nano-Medicine Healthcare	MHLW	1.416
Nanotechnology: Realization Project of Nanotechnology-Based Advanced Devices	METI	0.800

FIGURE 2 Japan's major R&D programs, 2005.
SOURCE: Presentation by David K. Kahaner, Asian Technology Information Program, "Japanese Technology Policy: Evolution and Current Initiatives," in Panel III of this volume.

tion of regional clusters around universities, while efforts are underway to raise at least 30 Japanese universities to the highest standards.

Taiwan's Transformation to a Knowledge-Intensive Economy

Taiwan's success reflects the power of state-led, market-oriented innovation policy. These policies have brought about a series of remarkable changes that has seen Taiwan grow from an agrarian economy with per capita GDP of $145 in 1951 to a modern industrial economy with GDP at $13,529 by 2004.[32] Notwithstanding this success, Taiwan now faces new challenges in the global economy. Executive Vice President of Taiwan's Industrial Technology Research Institute (IRTI), Hsin-Sen Chu, noted that Taiwan is in the process of transforming itself from a technology-intensive economy to a knowledge-intensive economy of the future.

While remaining market-oriented, government policies have, nonetheless, been instrumental in shaping Taiwan's industrial evolution. Policies enhancing Taiwan's industrial development include the founding of the Industrial Technol-

---

[32]For a review of Taiwan's technology policies, see Alice Amsden, *Beyond Late Development: Taiwan's Upgrading Policies,* Cambridge, MA: MIT Press, 2003.

ogy Research Institute (ITRI) in 1973, the establishment of the Hsinchu Science Park in 1980, and the Southern Taiwan Science Park in the 1990s. These initiatives have been supported by major investments in basic infrastructure development through ten large-scale public construction projects.

Describing ITRI's role in Taiwan's innovation system, Chu noted that ITRI's mission has been to engage in applied research and supply technical services to accelerate the industrial development of Taiwan. Specifically, ITRI develops key, compatible, forward-looking technologies to meet industrial needs and helps to strengthen Taiwan's industrial competitiveness. ITRI has 13 research units covering research in information and communications technology, advanced manufacturing and systems, biomedical technology, nanotechnology, materials and chemicals and energy and the environment. ITRI's role as a hub linking science parks, universities, and companies in Taiwan's north, central, and southern zones helps to link different parts of Taiwan's innovation ecosystem.

In 2004, ITRI employed 6,540 people, of whom 14 percent held doctorates. Of ITRI's $579 million budget in 2004, 52 percent came from the government, of which ITRI devoted about a quarter to the development of high-risk technologies. Another 40 percent of ITRI's revenue came from technology transfer to industry, with the remainder derived from its intellectual property. Describing ITRI's impact, Hsin-Sen Chu noted that, over 30 years, ITRI has helped guide Taiwan's transformation into the world's fourth-largest producer of IT hardware. In addition, he noted that Taiwanese firms now make up 73 percent of China's IT production.

Keeping in mind the competitive advantages of the mainland, Chu said that the Taiwanese policy makers see future opportunities in high-value manufacturing, novel applications and products, and knowledge-based service industries. To enhance the potential of Taiwan's national innovation system in this new era, the government is not only pursuing the creation of basic infrastructure and enhanced technological competency of Taiwanese firms, it is also promoting the development of a business environment that will promote stronger partnerships among industry, academic organizations, and industrial firms. An effective transition to capture the opportunities of the 21st century requires an adjustment of mindset, he added, and he further observed that there was an effort to move from optimization to exploration, from ordering work by single discipline to multidisciplinary integration, from conducting research in-house to collaboration and partnership, and from developing components to developing system solutions.

COMMERCIALIZING INNOVATIVE TECHNOLOGY IN THE UNITED STATES

In comparison to the preceding *tour de la table* of foreign initiatives, what role do innovation programs play in the United States? Notwithstanding the frequent U.S. rhetoric concerning the primacy of idealized markets, national policies in the

---

BOX B
Key Examples of U.S. Public-Private Partnerships[a]

1798—U.S. grant for production of muskets with interchangeable parts, to Eli Whitney, who founds first machine-tool industry
1842—Samuel Morse receives an award to demonstrate feasibility of telegraph
1919—RCA is founded on the initiative of U.S. Navy with commercial and military rationale
1969-1990s—U.S. investment in forerunners of the Internet
Present—U.S. investments in genomic/biomedical research

---

[a]This list was presented by Marc Stanley. See the proceedings section of this volume for a summary of Mr. Stanley's presentation. For a more detailed overview of the positive federal role, see National Research Council, *Government-Industry Partnerships for the Development of New Technologies: Summary Report*, Charles W. Wessner, ed. Washington, D.C., 2003, Chapter IV.

United States have long helped to foster American innovation—often decisively.[33] And recent public-private partnerships have been widely credited with reviving the U.S. semiconductor industry and the U.S. supercomputer industry.[34]

Indeed, such public-private partnerships have long played instrumental roles in developing new, game-changing industrial processes, products and services. (See Box B.) As Vernon Ruttan has observed, "Government has played an important role in the technology development and transfer in almost every U.S. industry that has become competitive on a global scale."[35]

A Comeback in Supercomputing

As a recent example of a U.S. public-private partnership, Kenneth Flamm of the University of Texas described the role of the superconducting partnership in reclaiming U.S. leadership in this strategic technology. Flamm began his narrative

---

[33]Contributions include the telegraph, the development and commercialization of the radio, aircraft engines and airframes, radar, nuclear power, satellite communications, the Global Positioning System, and, of course, the Internet. For a discussion of the federal government's role, see Irwin Lebow, *Information Highways and Byways*, New York: Institute of Electrical and Electronics Engineers, 1995.

[34]See National Research Council, *Securing the Future: Regional and National Programs to Support the Semiconductor Industry*, Charles W. Wessner, ed., Washington, D.C.: The National Academies Press, 2003.

[35]Vernon Ruttan, *Technology, Growth and Development: An Induced Innovation Perspective*. Oxford, UK: Oxford University Press, 2002, p. 301.

from the mid-1960s to the late 1970s, when the entire supercomputer industry basically resided in two American firms: Control Data and Cray. The Japanese, he said, entered the computer market only in the mid-1980s, initially producing IBM compatibles. However, a focused innovation policy initiated with the Fifth Generation Computer Project and the Super-speed Computer Project, helped Japanese producers to rapidly make significant inroads into the high-performance mainframe computer market.

Recognizing that superiority in information technology systems was essential to a qualitative advantage in defense systems, the Defense Advanced Research Projects Agency (DARPA) launched its Strategic Computing Initiative in the 1980s.[36] Although DARPA managers originally focused on custom components to build new computer architectures, they gradually switched their emphasis to methods of lashing together relatively inexpensive commodity processors into massively parallel systems. This effectively shifted the "terms of the battlefield" rather than meet directly the threat from "very, very well done" high-performance processors from Japan.

The result has been the renewed ascendancy of the U.S. supercomputer industry. The U.S. industry share of the top 500 machines sold has grown steadily while the Japanese share has been shrinking. This positive picture differed little, Flamm added, if looked at from the point of view of total computing capability. Furthermore, U.S. market share has been increasing not only worldwide, but in each individual region of the globe in terms of sales as well as computing capability.

Citing a finding of a recent National Academies study, Flamm noted that the government-industry partnership formed to develop alternative methodologies for designing and building supercomputers successfully transformed the nature of the supercomputer market over the past 10 years.[37] The policy implemented in the 1990s proved to be a huge success, even though the eventual outcome did not match the original plan. In closing, Flamm held up the resurgence of the U.S. supercomputing industry as "an example of a government-industry partnership in technology development that has yielded unforeseen but impressive results as an industrial outcome for the United States."

---

[36]The other response to the Japanese challenge of the 1980s took the form of trade-policy initiatives, one of those being an attempt to open up Japan's market through forcing procurement by its government of U.S. supercomputers. In addition, antidumping cases were filed in the United States in the mid to late 1990s.

[37]National Research Council, *Getting Up to Speed: The Future of Superconducting*, Susan L. Graham, Marc Snir, and Cynthia A. Patterson, eds., Washington, D.C.: The National Academies Press, 2005.

Early-Stage Funding and the Advanced Technology Program

Public-private partnerships can also represent a pragmatic institutional response to market failures in early-stage finance.[38] Although U.S. capital markets are relatively broad and deep, private investors often find the risk levels associated with investments in innovations that are still in their early stages to be too high and are therefore (understandably) reluctant to invest in unproven innovations. Even when private investors see manageable risk, they may not see ways to capture returns from their investment due to technology "leakage" or "spillovers" to other firms.

The Advanced Technology Program (ATP) is designed to address this challenge. As NIST's Marc Stanley noted at the symposium, of the roughly $20 billion in venture capital investments in 2004, only $375 million was available for the initial seed rounds. This is because most private equity investors prefer to invest in less risky later rounds of investment. These investments also tend to concentrate in a limited number of geographical regions.[39] This gap in investment at the seed and early stages is often called the "Valley of Death."[40] (See Figure 3.)

The mission of the ATP is to help bridge this valley between the research laboratory and the marketplace. To do so, ATP provides highly competitive awards, largely (about 70 percent) to small companies and to joint ventures designed to accelerate the development and dissemination of high-risk technologies with potential for broad-based economic benefits to the U.S. economy. (See Box C for a comparison of the ATP with Finland's Tekes Program.) ATP funding is not a "fire and forget" program. The awards to larger firms must be matched on a cost-share basis. They are closely monitored and can only be directed to technical research and not product development.

The program is entirely industry-driven. Companies, whether singly or jointly, conceive, propose, and execute all projects, often in collaboration with universities and federal laboratories. ATP support for project costs is limited in time and amount. Based on a rigorous merit-based competitive evaluation that

---

[38]National Research Council, *Government-Industry Partnerships for the Development of New Technologies: Summary Report*, Charles W. Wessner, ed., Washington, D.C.: The National Academies Press, 2003.

[39]As the venture capitalist David Morgenthaler has observed, "It does seem that early-stage help by governments in developing platform technologies and financing scientific discoveries is directed at exactly the areas where institutional venture capitalists cannot and will not go." See National Research Council, *The Advanced Technology Program: Assessing Outcomes*, Charles W. Wessner, ed., Washington, D.C.: National Academy Press, 2001.

[40]In his presentation, Canada's Peter Nicholson refers to an "orphaned domain" in the spectrum that ranges from basic research to commercial markets. In this domain, it is difficult to discern whether social returns or private returns on innovation are higher. This ambiguity leads to a lack of adequate funding. Similar to the concept of the Valley of Death, this orphaned domain is the focus of Canada's innovation programs.

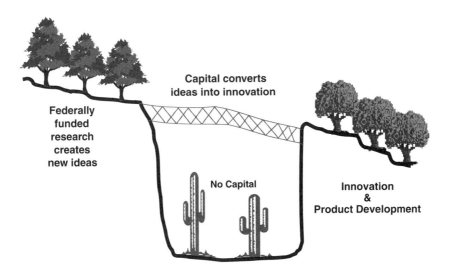

FIGURE 3 The Valley of Death.

admits less than 15 percent of applicants, single company awardees can receive up to $2 million for R&D activities, joint ventures considerably more.[41]

Stanley noted that ATP's selection process, monitoring, and follow-up on projects have been recognized by the National Academy of Sciences as being exceptional, adding that the program has demonstrated both the ability and the willingness to identify unsuccessful projects and, if necessary, end them. "You have to terminate companies that are not successfully doing what they say," he commented. "And then you should be able to speak not only of your successes but of your failures, because there are lessons to be learned from both." Based on a sample of 41 of the 736 projects it has funded, ATP analysis has calculated net societal benefits of $17 billion—representing a partial return on the $2.2 billion investment by the federal government over the life of the program.

Improving Technology Transfer from the National Laboratories

This policy ambivalence has affected the returns from U.S. laboratories as well. While the United States makes significant investments in its national labo-

---

[41]For an assessment of ATP, see National Research Council, *The Advanced Technology Program: Assessing Outcomes,*, op. cit. The NRC assessment found that ATP "is an effective federal partnership program" and that it is meeting its legislative goals in creating broad-based benefits to society, contributing to important social goals such as improved health diagnostics, and improving the efficiency and competitiveness of U.S. manufacturing. The NRC Committee conducting this assessment also offered a series of operational recommendations to make the program more effective (p. 87).

BOX C
Comparing the U.S. ATP Program with
Finland's Tekes Program

Several speakers at the conference compared the relative sizes of the United States Advanced Technology Program with Finland's similar Tekes program. ATP's impact at the cutting edge of new technologies is based on a relatively small annual budget. In 2005, the budget for ATP was $140 million, in the context of a $12 trillion economy and a population of 300 million. By comparison, Finland's Kotilainen noted that Tekes—a program that is similar to ATP—is financed at an annual level of around $550 million, supporting the innovation system of a nation of five million.

This relatively high level of expenditure reflects the strong consensus present in Finland regarding the need to support the technological enhancement of existing industries and to support the growth of promising new high-tech industries. Tekes awards for R&D effectively encourage partnerships between university researchers and small and large companies. Like the ATP, Tekes maintains a careful evaluation program that has recorded numerous success stories, with its early support for the research that contributed to the transformation of Nokia being perhaps the most notable example. The scale and scope of Tekes activity underscore the Finnish Government's and society's commitment to supporting the development and adoption of new technologies, particularly those subject to first mover advantage in order to capture the benefits of these innovations for the national economy.

In recent years, U.S. policy makers have been much more ambivalent about the appropriate role of government contributions to the development of new technologies, even as government supported technologies have transformed the economy.[a] Support for the Advanced Technology Program has been uneven and subject to the vicissitudes of the political process despite a positive assessment by the National Academies. It appears slated for elimination perhaps reflecting in part longstanding U.S. ambivalence about the appropriate role for government in encouraging innovation, as distinct from basic research.[b]

---

[a]See, for example, National Research Council, *Funding a Revolution: Government Support for Computing Research,* Washington, D.C.: National Academy Press, 1999, pp. 5-14 and passim.

[b]*Computer World,* "Bush May End Federal Tech Funding Program—Program for High-risk IT Projects is at High Risk of Elimination," February 12, 2006. The Office of Management and Budget provides this rationale for eliminating ATP: "Consistent with the Administration's emphasis on shifting resources to reflect changing needs, the 2006 Budget proposes to terminate the Advanced Technology Program. This proposal is consistent with the 2005 Consolidated Appropriations Act which did not provide funding for new awards. The Administration believes that other NIST programs are more effective and important in supporting the fundamental scientific understanding and technological needs of U.S.-based businesses, American workers, and the domestic economy." Executive Office of the President, *Budget of the United States Government Fiscal Year 2006,* accessed at *<http://www.whitehouse.gov/omb/budget/fy2006/budget.html>*.

BOX D
Spurring Technology Transfer from the National Laboratories

Pace VanDevender pointed out that the Department of Energy's (DoE) involvement in commercializing innovation began in 1980 with the *Stevenson-Wydler Act*, which established technology transfer as a mission for the federal laboratories, with a focus on disseminating non-classified information. This was followed by a series of acts that attempted to spur commercialization, beginning with the *1984 Trademark Clarification Act*, which gave the contractors that operated the laboratories licensing and royalty authority for the first time as an incentive to commercialize innovative ideas that were born in the laboratories. The *1986 Technology Transfer Act* then extended this responsibility to laboratory employees. Next, the *National Competitiveness Technology Transfer Act* of 1989 extended the technology transfer mission to the DoE's weapons laboratories. It also allowed the contractors that ran the laboratories to enter into cooperative research and development agreements (CRADA) so that they could enter partnerships with industry in cofunding further R&D. This was followed by the *1989 NIST Authorization Act*, which recognized intellectual property other than inventions that have been developed by CRADA, clarifying a legal uncertainty. Then, in 1995, the *National Technology Transfer Act* guaranteed to industry the ability to negotiate for rights to CRADA inventions and increased the royalty distribution that were placed on laboratory inventions, thereby increasing the motivation to invent.

---

ratories, the record of successful technology transfer to commercial applications has been relatively limited, according to Pace VanDevender of Sandia National Laboratories.[42] This modest outcome comes despite a long series of legislative experiments that have repeatedly sought to create the incentives needed to spur technology transfer from the national laboratories. (See Box D.)

How well have these technology transfer policies worked? According to Pace VanDevender, DoE transfer activities in 2004 included "some respectable numbers." (See Figure 4.) At the same time, he acknowledged that a single laboratory was responsible for approximately half of the 10,000 technology transfer initiatives that took place during that year; on a lab-to-lab basis, therefore, technology transfer activity has been "fairly modest." (He did not elaborate, however, on the reasons why this particular laboratory was relatively successful.)

---

[42]These investments are expected to grow substantially. The Department of Energy's (DoE) FY2007 budget requests of $4.1 billion for the DoE Office of Science is a $505 million (14.1 percent) increase over FY2006 funding. This budget puts DoE's Office of Science on the path to doubling its budget by FY2016. See Department of Energy Press Release of February 2, 2006, "Department Requests $4.1 Billion Investment as Part of the American Competitiveness Initiative: Funding to Support Basic Scientific Research."

• More than 10,000 technology transfer actions in FY 2004	• CRADAs	FY 2004
	— Active CRADAs	610
	— New CRADAs	157
	• Intellectual Property	
• Incorporates activities across DOE complex of 24 national labs/facilities	— Invention Disclosures	1,617
	— Patent Applications	661
	— Issued Patents	520
	• Licenses	
• Supports DOE/NNSA missions by enhancing lab capabilities and commercializing technologies	— All Active Licenses	4,345
	• New Licenses	616
	— Patent Licenses — Active	1,362
	— Other IP Licenses — Active	2,983
	— Income-Bearing Licenses	3,236
	• Active WFO/NFE Agreements	1,884
	• Active WFO/Other Fed. Agreements	2,782
	• User Facilities Agreements/ FY04	3,252

FIGURE 4 Technology transfer supplements the primary missions of each lab.
SOURCE: Presentation by J. Pace VanDevender, Sandia National Laboratories, "Sandia National Laboratories: DoE Labs and Industry Outreach," in this volume.

Also of concern, he noted, was a significant drop-off in CRADA activity after a rapid increase from 1992 to 1996, when federal matching funds were no longer available. Similarly, the growth of invention disclosures similarly hit a plateau in the late 1990s. Meanwhile the growth of patent applications and patents granted remained modest. (See Figure 5).[43]

Success at Sandia Science and Technology Park

In contrast to the CRADA, patent applications, and other tech-transfer vehicles whose growth has recently flattened, VanDevender noted that science and technology parks are emerging as a new thrust for the Department of Energy's technology-transfer efforts. The Sandia National Laboratories Science and Technology Park by its seventh year drew $167 million in investment (of which $146.6 million was private) and is still growing.[44] A campus-style 200-acre installation,

---

[43]An invention disclosure is a document which provides information about inventor(s), what was invented, circumstances leading to the invention, and facts concerning subsequent activities. See for example, Stanford University, *Inventions, Patents and Licensing: Research Policy Handbook*, Document 5.1, July 15, 1999.

[44]See National Research Council, *A Review of the Sandia Science and Technology Park Initiative*, Charles W. Wessner, ed., Washington, D.C.: National Academy Press, 1999. The Sandia report and consultative process it stimulated helped the Sandia National Laboratory's decision to establish an S&T park and helped to shape its structure.

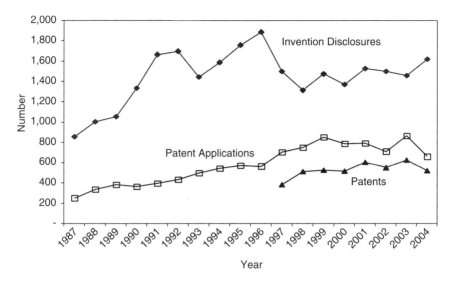

FIGURE 5 Invention disclosures and patents have plateaued under current policies and priorities.
SOURCE: Presentation by J. Pace VanDevender, Sandia National Laboratories, "Sandia National Laboratories: DoE Labs and Industry Outreach," in Panel IV of this volume.

Sandia Park by the spring of 2005 housed 19 organizations with 1,098 employees that occupied almost 500,000 square feet. Sandia and the park tenants enjoy a symbiotic relationship. Sandia National Laboratories provides redundant power and state-of-the-art connectivity to the park tenants and helps to accelerate city approval processes. Tenants in turn paid in $17 million to Sandia while acquiring contracts from the laboratory worth $85.6 million as of spring 2005. "The government, Sandia, and industry therefore benefit both ways" observed VanDevender.

Comparing DoE and ITRI Technology Transfer Models

Contrasting Taiwan's ITRI and DoE's technology transfer models, VanDevender said that ITRI was based on a single-purpose mission of technology development and commercialization with relationships, while the main mission of the DoE labs was "national security broadly writ." For ITRI, therefore, technology transfer was a dedicated mission, whereas for DoE it was a supplementary mission, not one central to the management's intent. The DoE labs received about ten times as much annual funding as ITRI, or $6 billion versus $600 million. But industrial contributions accounted for only about $60 million of the DoE labs' funding, or 1 percent, while around $200 million, or one-third, of ITRI's funding came from industry.

The DoE labs produced around 600 patents per year, half as many as ITRI's 1,200; this translated to 0.1 patent per $1 million for DoE against two patents per $1 million for ITRI, a yawning gap in the rate at which commercially valuable property transferred. But the gap in patents per industry dollar was far narrower, and the figure for the DoE labs was higher—about 10 patents per $1 million versus six patents per $1 million for ITRI—because at DoE industry was leveraging the huge U.S. investment in national security. But these two rates were called "very comparable" by VanDevender, "given the uncertainty in the value of those patents, [and] particularly since a whole lot more companies get spun off from ITRI than from DoE." Both models have their strengths and both were valuable, he concluded, suggesting that the comparison raised a question worth considering at the next stage of policy making: "whether [the United States] should reinvigorate a single-system kind of laboratory, perhaps much more like [what] ATP is doing with industry."

U.S. POLICIES IN COMPARATIVE PERSPECTIVE

There is strong international interest in national measures to attract and grow globally competitive, high-technology industries. Perhaps what is most striking is the range of mechanisms, the similarity of goals, and the very substantial resources devoted to building the infrastructure and technological capabilities—not for national security—but for national competitiveness in the global economy.

With some exceptions (e.g., small business award programs, such as SBIR and ATP), U.S. policy is not focused on the innovation process itself; resources are instead concentrated on particular research challenges and national security missions. As technologies evolve more rapidly, often in a multidisciplinary fashion, the importance of cross-disciplinary public-private partnerships seems likely to grow. This international comparative focus on innovation policy adds value by reviewing the range of these programs, underscoring the role institutions play in national policies and, implicitly, by reminding Americans of the accelerating competition for technological preeminence.

Despite this policy lacuna, the United States does possess great strengths. U.S. economic leadership rests on its large, integrated domestic consumer markets; deep and flexible capital markets (including risk capital); and deep and flexible labor markets. The United States also enjoys the advantages of an institutional framework—characterized by strong competition, the rule-of-law, and a willingness and ability to adapt new technologies that facilitate the rapid deployment of resources to take advantage of new opportunities.[45]

---

[45]Amar Bhidé, "Venturesome Consumption, Innovation and Globalization." Paper presented at the Centre on Capitalism & Society and CESifo Venice Summer Institute 2006, "Perspectives on the Performance of the Continent's Economies," July 21–22, 2006, held at Venice International University, San Servolo, Italy.

Box E
Research for Competitive Advantage

"Basic research has become part of the international competition of overall national strength."
Strategy document of the July Chinese State Council
Quoted by the New China News Agency, February 9, 2006

U.S. economic leadership is also supported by an entrepreneurial culture that encourages risk taking and tolerates failure. This entrepreneurial culture is reflected in and further reinforced by a supportive legal framework. This includes bankruptcy laws that do not excessively punish business failures and tax policies that permit successful entrepreneurs to retain significant portions of the wealth they generate. The legal regime is further reinforced by positive societal attitudes toward business success. This combination of mutually reinforcing attitudes and laws represents a unique competitive advantage for the United States, one that sets the U.S. apart. Calling this "a very special characteristic," Carl Dalhman noted that "many other countries really are trying to imitate" it, but with debatable success. In many other countries, if you take a risk and your business fails, the social and economic consequences can be dire and permanent. These positive attitudes toward entrepreneurship represent a major U.S. advantage in the risky world of new technologies and high-tech start-ups.

Resting on these foundations are a multiplicity of strong science and technology institutions, complemented (particularly in the post-war period) by strong investment in education. Another advantage, also noted by Dalhman, is that the United States is home to more multinational corporations than any other country.

As many of the conference speakers made clear, directly or indirectly, the environment in which the U.S. economy is competing has become much more competitive.[46] Many countries are now investing heavily in R&D, in education, and in science and technology infrastructure, often with a focus on specific technologies for the market. The U.S. advantage in terms of multinationals, with their benefits of expertise, integration, and market power, is also less preeminent. Other countries are now hosts to significant global corporations, not only in traditional areas (e.g., Europe) but increasingly in Asian countries such as Korea, Taiwan, India, and China.

---

[46]See also NAS/NAE/IOM, *Rising Above the Gathering Storm: Energizing and Employing America for a Brighter Economic Future*, op. cit.

As Sandia's Pace Vandevender emphasized at the conference, participants in the global economy recognize the importance of dedicated institutions. New institutions such as ITRI and Tekes have made Taiwan and Finland formidable competitors in important markets and laid the foundation for future strength. Similarly, U.S. strengths in the availability and diversity of early-stage capital, while still unsurpassed, are nonetheless being challenged. Where other countries cannot emulate the private risk taking that characterizes early-stage finance in the United States, they are taking measures to provide publicly supported capital and incentive schemes designed to blend private and public funds as a means of reducing risk and encouraging investment.[47]

Even the traditional U.S. strengths of a large, unencumbered domestic market, while not yet matched, are no longer as unique. Emerging economic arrangements—such as the European Union and ASEAN as well as large economies of emerging nations like China and India—have the potential to counterbalance U.S. economies of scale in the long term. At the same time, a strategic approach that focuses on the ability of U.S. firms to access other national markets, build cooperative relationships, and seek out expertise in a way that benefits both the United States as well as its global partners is required for continued U.S. leadership. A crucial condition for U.S. competitiveness is the extent to which federal and state governments invest in a robust S&T infrastructure and in effective programs to ensure supplies of scientists and engineering graduates and to facilitate the transition of research to the market.

Common Challenges, Diverse Approaches

The intense competition which characterizes the global economy has placed a premium on the capacity to innovate. Innovative companies are able to provide attractive new products that meet or create market demand. Companies that benefit from a supportive national innovation policies are able to compete more effectively. They can draw on a steady stream of well-trained graduates, increasingly with practical experience, and they benefit from supportive financing (e.g., innovation awards) that enable companies to convert the fruits of research to new welfare-enhancing products.

The common challenge for most participants in the global economy is the need to capitalize on their intellectual assets, converting government funded research into the innovative technologies and processes that generate improved welfare, create international competitiveness, and create wealth for their citizens. It is, perhaps, exceptional that countries as diverse as China, India, Taiwan, Japan, Germany, France, Finland, Canada, and the United States are all devoting substantial policy attention to the transition of research into products and processes

---

[47]See, for example, the presentations by Stephan Kuhlman and Finland's Heikki Kotilainen in Panel II of this volume.

of the future. What is equally remarkable is that, while the challenge is similar, the mechanisms and instruments adopted to encourage this transition show very considerable variation, albeit with some common features. The basic goal of this conference was to bring practitioners and analysts together to discuss the common goals and the diverse measures taken to achieve them.

The challenges of the twenty-first century point to the need to reexamine the policies supporting and building interconnections within the U.S. innovation ecosystem. As described in the conference proceedings that are summarized in the next chapter, the many foreign programs presented at this conference provide graphic evidence of the scope and scale of national efforts to enhance their national prospects in the global economy. The strong cooperative element of the conference also merits emphasis. The conference deliberations underscored the opportunity and indeed the need to learn best practice from the many national experiments underway.

II
PROCEEDINGS

Welcome Remarks

Charles W. Wessner
National Research Council

Dr. Wessner welcomed symposium participants to what he said promised to be an intense discussion of the innovation policies of a diverse group of countries with a focus on the mechanisms used to help facilitate the innovation process. He observed that those in attendance, many of whom were intimately involved in the innovation process, knew the day's topic to be complex and, at times, to be the subject of proposals that were ideological or simplistic. Because many countries around the world have adopted effective policies, it had also become increasingly urgent: Realization was growing in Washington, as it had in many other world capitals, that innovation and the mechanisms facilitating it are a key element in national growth and national competitiveness.

In fact, the entire world is focused on how to deliver the fruits of research through products and processes that both enhance welfare and generate wealth. In the STEP Board's work with other countries it had become clear, Dr. Wessner said, that the problems and challenges facing India, the People's Republic of China, Canada, Finland, Germany, and the United States were essentially identical, something without precedent in his own public life.

How do we capitalize our investments in research? How do we generate the type of students and the type of output from our universities that will help our economies to grow and to meet the challenges of the environment, of health care, and of providing a better life for our children? To discuss such issues, presenters had traveled to Washington from the four corners of the Earth. Dr. Wessner extended special thanks to Stefan Kuhlmann of the Fraunhofer Gesellschaft in Germany, to Peter Nicholson from the Office of the Prime Minister of Canada, and to Hsin-Sen Chu of ITRI, Taiwan. He also expressed particular appreciation to the IBM representative who was in attendance; to Intel; to Sandia National Laboratories; and to the National Institute of Standards and Technology (NIST), whose support made the symposium possible.

FOCUS ON THE "NATIONAL INNOVATION ECOSYSTEM"

The day's main focus, would be on how to link together universities, laboratories, and the private sector—both large companies and small—in an effective system of national innovation. A term for this, "national innovation ecosystem," had emerged from previous work by the STEP Board, which was pleased to see that the concept had been picked up by the President's Council of Advisors on Science and Technology (PCAST) and the Council on Competitiveness. Key to the formation and effectiveness of a national innovation ecosystem are what Dr. Wessner called intermediating institutions: the institutions, and with them the mechanisms, that can help bridge the diverse institutions that contribute to an innovation ecosystem.

Even as he acknowledged the challenge of creating the necessary linkages, he counseled that attention be directed toward incentives. "I am always disheartened when I hear someone in France say, 'We must reform the university system,'" he recounted. "It always reminds me of the Battle of Verdun: a very long, bloody struggle that in the end accomplished little other than the bloodshed." While it is difficult to transform organizations by fiat, "we like to think," he said, "that it is possible with appropriate incentives." The pleasure of the day's endeavor, said Dr. Wessner, would be in the opportunity to listen to a number of experts in this field engaging in what were referred to as "interesting experiments" by the vice-chair of the STEP Board's Steering Committee for Government-Industry for the Development of New Technologies, William Spencer.

Introducing Dr. Spencer, Dr. Wessner recalled the vital role that he had played in leading Sematech at a time when, in the opinion of many economists, the U.S. semiconductor industry was on the ropes and destined to become a marginal player in the world market. But being too busy, as it appeared, to read such prognostications, Dr. Spencer and the industry had "just kept going," with a program to cooperatively improve product quality and output. As a result of the Sematech consortium, the trade agreement, and much hard work and inventiveness, the U.S. semiconductor industry is now ranked first in the world.[1] Reiterating the importance of Dr. Spencer's contribution to the industry's recovery and the industry's impact on U.S. productivity and competitiveness, he asked Dr. Spencer to come to the podium.

---

[1]For a description of the factors contributing to the resurgence of the U.S. semiconductor industry, see National Research Council, *Securing the Future: Regional and National Programs to Support the Semiconductor Industry,* Charles W. Wessner, ed., Washington, D.C.: The National Academies Press, 2003.

Opening Remarks

William J. Spencer
SEMATECH, retired

Dr. Spencer expressed his appreciation to Dr. Wessner and joined him in welcoming the participants to the day's program and in thanking its sponsors. He singled out for thanks both Taffy Kingscott of IBM and Marc Stanley, the head of the Advanced Technology Program at NIST, which is housed at the U.S. Department of Commerce, and acknowledged the contributions of Intel and Sandia. He introduced the symposium, as one of a series that will be organized by the STEP Board over several years. Dr. Spencer posited that a poll of those present would show relatively strong agreement that technology plays an important role in economic growth irrespective of region. He further suggested that there would be fairly uniform agreement that technology is leading to better quality of life, although he noted such dissenting voices as that of Bill McKibben, whose *Enough: Staying Human in an Engineered Age* raised questions about germ-cell engineering, and of Bill Joy, the former chief scientist at Sun Microsystems, who had expressed concern about nanorobotics.[2]

FUNDING INNOVATION: PRIVATE OR PUBLIC?

There would likely be a divergence of opinion, however, if the subject was focused on who should fund science and technology. When it came to long-term research, most of us would probably agree that government should play a major role whether the research is "curiosity driven" or "problem driven"—that is,

---

[2]See William McKibben, *Enough: Staying Human in an Engineered Age,* New York: Henry Holt & Co., 2003, and William Joy, "Why the Future Does Not Need Us," *Wired*, 8.04, April 2000.

whether it dealt with things like string theory or quantum gravity that explain the universe, or with such questions as why some individuals have genes that are susceptible to disease and others do not. But when it comes to who should fund the innovations coming out of this long-term research, said Dr. Spencer, "I suspect the opinions would be more divergent." The purpose of the day's symposium, and of any follow-on meetings for which STEP might obtain the resources, was to try to gather facts on how innovation and technology transfer were being funded in the various economic regions, and in particular on the roles of private and of public funding.

Pointing out that any future meetings in the series on Comparative Innovation Policy would be principally organized by a steering committee, he recognized the members of that panel who were in attendance—Mark Myers, Lonnie Edelheit, Alan William Wolff, Alice Amsden, and Kenneth Flamm.[3]

Dr. Spencer then turned the microphone over to Bradley Knox, a member of the staff of the House Committee on Small Business, who chaired the opening session.

---

[3]See the complete committee list in the front matter of this report.

An Overview of the Global Challenge

Moderator:
Bradley Knox
House Committee on Small Business

Speaking on behalf of the House Small Business Committee, Mr. Knox expressed excitement that the symposium was taking place and gratitude to the committee and Dr. Wessner for bringing to the fore the matters it was to take up. He then introduced Carl Dahlman of Georgetown University, saying the audience was certain to benefit from his experience. Dahlman's previous career at the World Bank has spanned a quarter-century.

THE INNOVATION CHALLENGE:
DRIVERS OF GROWTH IN CHINA AND INDIA

Carl J. Dahlman
Georgetown University and The World Bank, retired

Dr. Dahlman said that he would rapidly draw a broad sketch of the world based on work that he has engaged in for several years. Following this, he stated that he would sketch out the strengths of the United States, the European Union, and major nations in Asia, highlighting some of the lessons their various experiences might provide. Next would be a discussion of some key drivers of growth and competitiveness, notably education and innovation—particularly with respect

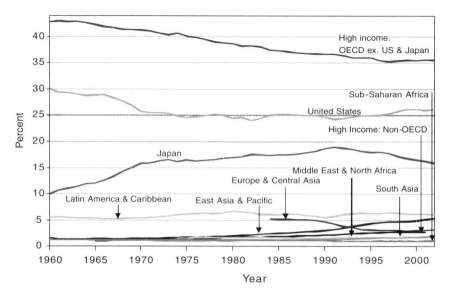

FIGURE 1 Shares of world GDP, 1960-2002.

to China and India. Finally, he noted that he would make some hypothetical projections intended to stimulate discussion.

Beginning with a chart showing changes in the relative size of different world economic groups between 1960 and 2002 (Figure 1[4]), Dr. Dahlman pointed out that the United States' share of global GDP fell from 30 percent at the start of that period to around 27 percent by its end. Meanwhile, Japan's share rose, particularly during the 1970s and 1980s, drawing the attention of its economic competitors. As the 1990s began, however, Japan "got stuck," said Dr. Dahlman, while the United States, whose share of worldwide GDP fell below 25 percent by 1991, began an economic recovery. In the meantime, the share of global GDP of the remaining OECD countries had shrunk more than that of the United States. Significantly, he added that the only region of the globe to record a continuous increase in its share of world GDP was East Asia (excluding Japan).

Rapid Growth of Chinese, Indian Economies

Dr. Dahlman's next graph, covering the period 1990-2002, assigned per capita GDP to the horizontal axis and average annual growth rate per capita to the

---

[4]Throughout this volume all dollars are U.S. unless otherwise indicated.

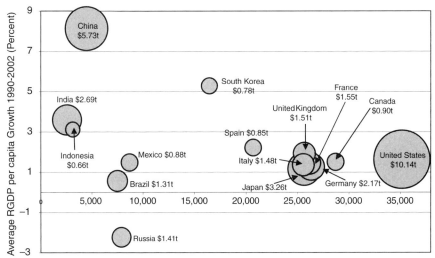

FIGURE 2 Fifteen largest economies (GDP 2002).
SOURCE: The World Bank.

vertical axis (Figure 2). Individual countries were represented by circles whose size corresponded to the size of their national economy measured in terms of purchasing-power parity (PPP) rather than in terms of nominal exchange rates (the latter, he asserted, may not always be a reliable guide.) While admitting the use of PPP is not beyond being questioned, Dr. Dahlman said his purpose was to emphasize the speed of China's growth: Measured in PPP, China's economy has for quite some time been second in size to that of the United States. According to his chart, Japan's economy was third largest and India's fourth largest, ahead of Germany's in fifth place. Not only were China's and India's economies "becoming big in the global sense," both were growing very rapidly; India was growing at 7 percent to 8 percent per year.

Dr. Dahlman then compared the various economic groups' shares of global GDP between 1990 and 2002 based on current U.S. dollars and on current PPP. Based on current dollars, the United States ranks number one among the world's economic groups, followed by the EU. After the EU come East Asia and the Pacific (EAP), grouping the Asian Pacific economies including Japan, and the group of the Big Six developing countries: China, India, Brazil, Russia, Indonesia, and Mexico. But using purchasing-power parity to measure GDP, the Big Six constitute the largest economic group, followed by the East Asia-Pacific group, the EU 25, the United States, and, finally, the EU 15.

The Foundation of U.S. Preeminence

Dr. Dahlman then listed reasons for the preeminent economic position of the United States, which by itself accounted for more than one-quarter of the world's GDP:

- very large, integrated domestic markets;
- an economic institutional framework facilitating rapid deployment and restructuring to take advantage of new opportunities;
  - strong competition;
  - a deep and flexible capital market (including risk capital);
  - a deep and flexible labor market;
  - good rule of law;
  - very strong science and technology institutions; and
  - very flexible managerial organizational structures.

Among additional advantages he cited was a very strong investment in education, and particularly in higher education; this he attributed in part to the post-World War II GI Bill, which has moved the United States "ahead of everybody else." On top of this, the nation boasted strong infrastructures both for research and for information and communications; was home to more multinational corporations than any other country; enjoyed superior military strength as the sole remaining superpower; and benefited from an entrepreneurial culture. Calling this last "a very special characteristic," he noted that "many other countries really are trying to imitate" it, but with debatable success. What was "unique" about the United States, he observed, was that "when you take risk [and] you fail honestly, that's even a good thing" in the eyes of Americans, whereas elsewhere if "you take a risk [and] you fail, you're out."

Other Nations "Catching Up Fast"

Turning to the challenges that the United States is facing, he pointed out that "other countries are catching up fast." They are investing heavily in R&D, in education, and in science and technology infrastructure. Large multinationals have been multiplying, not only in Europe but in the developing countries of Asia—Korea, Taiwan, and China—as well. And one consequence of the growth of these very large, global corporations was that "there's some confusion on 'Who is us?'" Dr. Dahlman stated, posing the question: "Is Nokia Finnish, or is it now a 'global corporation'—and what does that mean?"

He included among other challenges to United States preeminence:

- emerging regional economic arrangements, such as the EU and ASEAN, with the potential to counterbalance U.S. economies of scale;

- ongoing competition for global market share in PPP terms, which would depend on the United States' performance relative to that of China and Japan;
- "gigantic" fiscal and trade imbalances, a "very strong vulnerability" of the kind that normally raises red flags at the World Bank, which the United States could sustain only because it owned the world's reserve currency; and
- some neglect of existing scientific and technical infrastructure in the face of vigorous efforts by other nations in this area.

Regional Economic Blocs

European Union

The EU was showing more success in creating large markets for trade in goods than for trade in services. Hallmarks of progress include a significant integration to a single currency, increased regional stability, increased policy coordination, and the incorporation of ten new countries as of May 1, 2004.

But the significant challenges rooted in the EU's low productivity growth and low economic growth remained. This challenge was compounded by EU expansion, as the group of countries across which policies needed to be coordinated became much larger. Making productivity and growth rise would require far more flexibility in the EU's economic institutional regime, in its labor markets, in its capital markets, and in many rules and regulations at the local level. It would also require strengthening those educational fields most closely tied to research and innovation; despite having instituted programs with that objective, the EU was still lagging the United States in the area. The EU was also facing a major structural impediment in the aging of its population, which would prove a much larger burden there than in the United States, according to Dr. Dahlman. The dependence of a greatly increased number of retirees on a much smaller workforce would be "a big drag" on Europe's economy. Finally, the adjustment necessary to accommodate new entrants, including modernization of institutional and regulatory structures, would have to continue in the context of competitive pressures heightened by East Asia's joining the United States as the EU's rival for global markets.

East Asia

This region's market, although quite large to begin with, is the fastest growing regional market in the world. Intraregional trade, which accounts for half of its overall volume, is on the rise. Still, East Asia's markets are not fully integrated and East Asian economies remain very dependent on the U.S. and European markets.

At the same time, Dr. Dahlman said, "some critical mass" was reached in R&D and human capital in an area covering not just Japan but now China, India,

Korea, and Taiwan as well. Investment in education, and with it educational attainment, has been "tremendous." South Korea, for example, has rapidly transformed itself from a very poor country to one whose workforce in 2000 ranked third in the world in percentage of college-educated individuals, behind only the United States and the Netherlands. In fact, Korea could now be considered to have excess capacity in higher education. For its part, China has been ramping up its giant system of higher education by 50 percent per year for the past 5 years; as of 2004, it had more people in higher education than did the United States in absolute numbers. Several East Asian nations have undertaken rapid increases in their R&D efforts as well.

Korea and Taiwan in particular—but also Japan and, more recently, China— are investing very heavily in the information and communications technology (ICT) revolution, "rid[ing] this wave very well and [going] quite far with that." Meanwhile, East Asia has become home to a growing number of multinationals that have the ability, Dr. Dahlman observed, to "go out and compete with the big boys, do strategic alliances, and cross-subsidize from cash cows to the new areas."

The state's role in coordinating development strategy has been enlarged, since most of these countries see science and technology (S&T) as a key to future growth and competitiveness. The central focus for future growth is placed on S&T in the eleventh Chinese 5-year plan—released in March 2005, a month before this symposium—but this has in fact been the reality for a long time. The East Asian countries have placed emphasis on being fast followers and established a record of being very quick in that role. More recently, however, Korea and India have joined Japan in efforts to be on innovation's cutting edge, making significant investments and pushing back the frontier of knowledge.

Challenges for East Asia

East Asian nations nonetheless face significant challenges. They remain less developed than the United States, possess more rigid institutional structures, and continue to be dependent on the U.S. market. "This is a big risk for the whole world," Dr. Dahlman warned, "because they're serving as the ATMs for the U.S. deficits." Doing so had allowed them to prevent their currencies from appreciating, thereby keeping them competitive and ensuring their access to the U.S. market. China, for example, benefits from the dollar's devaluation with respect to the euro. But he characterized the situation as "a very delicate balance," asserting that it is impossible to predict what might happen in the event of major disruptions. Whenever authorities in these East Asian nations talk about moving from U.S. treasury bonds into euros or any other currencies, the markets become extremely worried.

In addition, fragmentation is a problem in East Asia, not just within the regional market but also within the various nations. China and India, Dr. Dahlman commented, are marked by a "tremendous lack of integration of the domestic markets." These countries' service sectors are less developed than that of the United States, as are their educational and innovation systems, although they were building those very rapidly. Moreover, with the exception of Japan and Korea, the East Asian nations have less developed value chains, marketing, and distribution networks. Again, however, China was beginning to build global value chains with its own brand names. Finally, allowing for variation from country to country, the region's culture was somewhat less entrepreneurial and risk-taking than that of the United States.[5]

Growth and Competitiveness: Key Drivers

Dr. Dahlman then described the factors that prompt what he called "a renewed interest in growth" over the previous decade, running through them quickly because he felt the attendees were already familiar with them. The ICT revolution had underlined the importance of knowledge as a major competitive element in the "New Economy," in which an increase in high-tech products' share of exports is increasingly accompanied by managerial and organizational changes. During this transition, macro-level evidence of changes have appeared in both the patterns and nature of growth in the industrialized world: There had been a reversal of a previous trend toward convergence of per capita income among OECD countries, and the growth of the U.S. economy between 1995 and 2002 was surprisingly strong.

The overriding development, however, is that globalization continues to increase. Over the previous 12 years, the share of worldwide imports and exports relative to total world outputs, or "world GDP," has increased from 38 percent to 52 percent, a fact that Dr. Dahlman attributed to a rapid reduction of transportation and communications costs. Contributing significantly to this trend are the biggest developers and disseminators of applied technology, the multinational or transnational corporations: The value-added by their production, in their home countries and by their affiliates abroad combined, accounted for 27 percent of global GDP. That figure, in fact, represents "an underestimate of their influence," he said, "because they have very long supply chains and also the forward linkages, marketing, distribution, [and] customer service." In addition, these corporations continue to be "very important in terms of strategy."

Changing Nature of Competitiveness

Traditionally, economists have viewed competitiveness as a function of capital and labor costs, the cost of inputs and infrastructure services, the general busi-

---

[5]Ashok Mody and Carl Dahlman, "Performance and Potential of Information Technology: An International Perspective," *World Development* 20(12):1703-1719.

ness environment, and strength in technology and management. This may have been appropriate to a more static system, but the current economy, stirred by the ICT revolution and by the volume of new knowledge being created and disseminated, is very dynamic. Competitiveness is coming to be based more and more on the ability to keep up with rapid technological and organizational advances, which affects the ability to redeploy resources both at the country level and at the firm or organizational level.

In this context, the importance of flexibility in labor and capital markets, and of social safety nets that would catch those people falling between the cracks, is increasing. The quality and skills of the labor force are assuming greater weight as well: As the half-life of knowledge has grown shorter, workers' level of formal education has grown relatively less important, and developing mechanisms for lifelong learning are relatively more important. "We are moving almost to just-in-time learning," observed Dr. Dahlman, stressing the need for systems capable of the rapid dissemination of knowledge "according to whatever is relevant for whatever institution or sector we're in." Another source of advantage resided in gigantic systems, such as those used by "the Wal-Marts," that are "very integrated with what the customer wants all the way to supply chains, massive economies of scale, and logistics." As shown by containerization's importance in permitting globalization through reducing transportation costs, it was not only "hard technology" that matters. The ability to make effective use of information technologies to lower transaction costs across the board also looms quite large.

The Knowledge Economy's Four Components

As a result, countries are placing added emphasis on improving their basic economic structure and business environment, their education skills, their innovation system, and their information infrastructure.

Dr. Dahlman then projected a graph (Figure 3) ranking world nations for innovation on a population-weighted basis as measured by three variables that, he said, had to be "very crude" because of the large number of countries included: scientists and engineers engaged in R&D per million population, scientific and technical publications, and U.S. patents. Each axis presents a rank ordering of all the countries in the world, but at a different moment: The horizontal shows the countries' positions in 1995, and the vertical shows their positions in 2002; the most advanced countries appear at the top right. Of importance is not only a country's relative position along the diagonal but also whether it is plotted above or below the diagonal: In the first case its position would have improved between 1995 and 2002, while in the second case it would have deteriorated. The graph indicates, therefore, that the United States has been stable, but that Korea, Brazil, China, and India have moved significantly in the direction of improvement.

A second graph (Figure 4), which differs only in that it is based on the countries' absolute size rather than population, provides a radically revised pic-

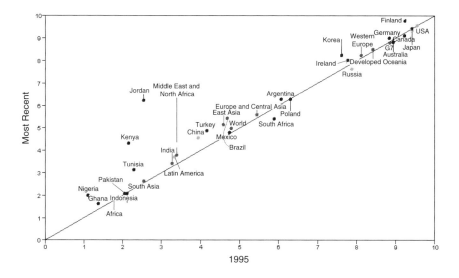

FIGURE 3 Innovation—Weighted by population.

ture. On this chart, Russia, Brazil, India, and China are found at the upper right, competing with the OECD nations. The reason, according to Dr. Dahlman, is that when it comes to knowledge—which, once produced, can be used without being consumed—critical mass matters.

China's Rapid Rise in R&D Spending

A third graph (Figure 5), placing R&D expense as share of GDP on the horizontal axis and scientists and engineers per one million of population on the vertical axis, presents a comparison of national efforts in R&D as of 2002 in terms of PPP. It depicts a clearly dominant United States, followed by Japan, with Germany, France, the UK, and China bunched behind the two leaders. But China has moved very quickly into third place, having increased its R&D investment from 0.6 percent of GDP in 2002 to 1.3 percent in 2003, a jump of 50 percent or, in PPP, 70 percent. It now boasts not only the world's third-largest R&D expenditure, but also its third-largest scientific and technical engineering work force focusing on R&D.

While acknowledging that there was no need to convince his present audience of innovation's growing importance for competitiveness, Dr. Dahlman emphasized that "innovation is not just about R&D [but also] about making effective use of experience to improve things," which can include imitating, copying, or buying

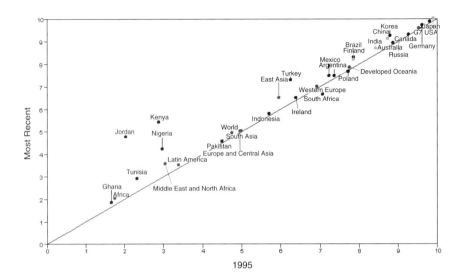

FIGURE 4 Innovation—Unweighted by population.

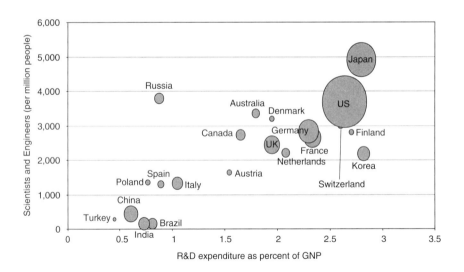

FIGURE 5 Global R&D effort in comparative perspective (PPP, 1996).

	Acquiring	Creating	Disseminating
Catch-Up	Most critical: -a lot of knowledge in pubic domain -also large stock to be purchased Therefore need good global scanning and acquisition ability	Less relevant or feasible, but still need R&D capability to acquire and adapt Critical to focus limited R&D efforts on most critical needs	Very important: -extension services -technical information -metrology, standards, testing and quality control -specialized suppliers -growth of most efficient firms
Countries Nearer Frontier or with Large Critical R&D Mass	Continue tapping global knowledge -FDI/licensing -Strategic alliances -foreign R&D as antennas to tap knowledge	Refocus public efforts on commercially relevant research Strengthen IPRs Increase private R&D efforts	Dissemination efforts continue to be critical But also need to commercialize knowledge -technology transfer offices -tech parks/spin-offs -cluster development

FIGURE 6 Differentiated strategies for innovation.

improvements made by others. It is valuable, he stated, to distinguish between innovation as "pushing back the global frontier" and as introducing knowledge, in the form of a development or application, into a local context. Being able to tap what is available from others was important because the stock of knowledge was moving so quickly and because nobody can enjoy an absolute dominance over the domain. For countries or sectors not yet at the global frontier, acquisition—whether through trade, foreign investment, or technology transfer—and adaptation of existing knowledge are paramount. For those closer to the frontier, pushing it back is what counts. Dr. Dahlman offered a chart sketching an innovations strategy appropriate to each of the stages (Figure 6).

Innovation's Engines, Private and Public

Naming the innovation system's key actors, Dr. Dahlman noted that multinational corporations (MNCs), as "the main generators and disseminators of technology," have been its engine in the private sector. He underlined the utility of determining how MNCs integrate their efforts with national policies, and the utility to other companies of learning how to link up with them. In national systems the key players are government research laboratories; universities; the enterprise sector, comprising not only established companies large and small but also startups; government innovation programs; and public-private partnerships.

These last are gaining in importance because, he said, "we're moving to some areas even where you have club goods and clusters."[6] He urged promoting greater interaction among the key players through such mechanisms as joint research grants, innovation awards, consortia such as Sematech, and other programs that rotate scientists and engineers, and government procurement officials.

There is also a need to generate more effective output from all these players, something that depends not on competitive pressure alone but, to a great extent, on the incentives extended to institutions and individuals. The "fine tuning" of these incentives, he said, "is critical." Science and technology policy, depending on the degree of backing it receives from government, has the potential to make this infrastructure work better through coordinating activities, administering public awards for innovation, and supporting high-tech parks, incubators, and technology transfer centers at universities.

While noting that all nations, faced with these issues, have been trying to devise policies, Dr. Dahlman singled out Finland for special commendation: Even though it had a "tiny" economy and a population of only five million, that country had put into place a very good system for developing the kinds of institutions that facilitate such coordination and linkages. The last need he mentioned was for mechanisms of evaluation and monitoring that would provide clear definition of goals, stipulate what was to be measured and how, and assess impact—all indispensable to learning how to use resources more efficiently.

Enumerating China's Economic Strengths[7]

To begin a discussion of the two top competitors emerging from the developing world, China and India, Dr. Dahlman observed that the former, by growing at about 8 to 10 percent per year for the previous four decades, has established the record for the fastest economic growth over the longest period of time for any country in the world. Its "gigantic" internal market affords it a very important strategic advantage in negotiating externally, as evidenced by the fact that foreign interests competing to invest in China had been willing "to bring not the second- or third-weight technology but the very best" for application in their operations there. Based on personal observations conveyed to him by a friend, he reported that a Motorola plant in China had two production lines. The line using more traditional technology made goods for export to the United States, while

---

[6]A club good is an impure public good whose benefits are excludable (nonmembers can be denied access to the good) but partially non-rival (a club member's enjoyment of a good only partially diminishes another's enjoyment.) James M. Buchanan first developed the theory of clubs in 1963. Reprinted in James M. Buchanan, "An Economic Theory of Clubs," in *Economics: Between a Predictive Science and Moral Philosophy,* College Station, TX: Texas A&M University Press, 1987.

[7]Carl Dahlman and Jean-Eric Aubert, *China and the Knowledge Economy: Seizing the 21st Century,* Washington, D.C.: The World Bank, 2001.

the line using more modern technology made goods for internal consumption. "That's what a big market does for you," he declared.

He then offered a catalogue of China's other economic strengths:

- It has a very high savings and investment rate, which, at about 40 percent, contrasts with 20-plus percent in most of the rest of the world.
- It is excellent at tapping into global knowledge through direct foreign investment and the Chinese Diaspora, the latter providing China and Taiwan a "fantastic global network . . . that is very hard to replicate."
- It is becoming the world's manufacturing base.
- It has a very large supply of excess labor in the agricultural sector, some 150-200 million people, which could continue to provide it a labor-cost advantage.
- It is nonetheless moving up the technology value chain very rapidly to become an exporter of far more than low-cost, labor-intensive goods.
- Its "fantastic" export-trade logistics, combined with economies of scale, make it "cheaper to ship from most ports in China to the U.S. than from most parts of Mexico to the U.S." despite the greater distance involved.
- It has achieved critical mass in R&D, which it is beginning to deploy in a highly focused effort to increase its competitiveness.
- It is making very strong investments in education and training.
- Its government has a very strong sense of national purpose, something that "helps to coordinate things, although it creates some other kinds of problems."

Lessons from China's Experience

Outlining lessons to be drawn from China's experience, Dr. Dahlman pointed to its demonstration of the "importance of the nation-state" not only in developing long-term plans and visions but also in providing a stable macroeconomic framework. He underlined what he called the "tremendous pragmatism" exhibited by the Chinese: "Although it is supposed to be a communist system, they have stock incentive plans in the research institutes." Similarly remarkable, he noted, was that one-third to one-half of the cost of higher education was paid by the students through tuition. While the Chinese have been focusing on technology and education for the previous two decades, the policies currently in development are more coordinated than those that had preceded them. "They are just really revving this up even more," he commented.

Yet more lessons might be found in China's conduct of its external relations. It has turned to "tremendous advantage" the realignment necessitated by its integration into the world economy. "Joining WTO was a risky move on their part, but it's given them lots of benefits," he said, counting among them not only the country's dominance of world textile markets but also the pressure that had come upon its domestic system to improve. China has been very effective at using foreign investment, first to move up the technology ladder, then to create home companies that have shown their strength in everything from competing with Cisco to buying IBM's notebook-computer capability. "And," he predicted, "that's just the beginning." Strong investment in human capital complemented by effective use of the Chinese Diaspora, whether in acquiring technology or gaining access to markets, filled out the picture.

India "Poised To Do a China"[8]

For its part, India has seen its annual growth rate rise from the 2 to 3 percent that was traditional prior to the past decade through the 5- to 6-percent level to around 8 percent. It was, in Dr. Dahlman's words, "poised to do a China," held back only by its own internal constraints. Chief among these was the surfeit of bureaucracy stifling a flair for entrepreneurship that is nonetheless very strong, as could be seen in the United States, particularly in California, and in linkages back to the home market from overseas. But the country has a critical mass of capable, highly trained scientists and engineers, most notably in the chemical and software fields. In addition to playing a prominent role in the outsourcing of business processes, it is becoming very attractive to multinationals as a place to conduct R&D.

In fact, because of India's tremendous cost advantage in human capital, more foreign firms have located large R&D facilities there than in China. Companies such as Wipro are increasingly performing contract research in India on behalf of multinationals—and in pharmaceuticals as well as in ICT, a sign of the "tremendous strength being built up there." The country has relatively deep financial markets compared to other developing countries, and, under the pressure of China's liberalization, is finally beginning to look not just internally but also outside. It is also seeking strategic alliances, aided by success in capitalizing on its own Diaspora for access to information and markets.

India's Human Capital Investment Pays Off

One of the main lessons to be drawn from the Indian experience is the significance of the long term: The investments in high-level human capital that were now beginning to pay off for India were made as far back as Prime Minister Nehru's time in the 1950s through mid1960s. The Indian Institute of Technology and Indian Institute of Management, world-class institutions that accepted only about 2 percent of applicants, have helped build a truly gigantic skill pool. "If they were to rev this up, they could begin almost to dictate the wages for anything you can digitize," said Dr. Dahlman, admitting this was an exaggeration calcu-

---

[8]Carl Dahlman and Anuja Utz, *India and the Knowledge Economy: Leveraging Strengths and Opportunities*, Washington, D.C.: The World Bank, 2005.

lated to be "provocative." Still, he observed, "they have a large mass of people that can move and, in economics terms, they have the production-functional educational system. They just have to expand it."

India could be expected to provide further positive lessons depending on how it handles other pivotal aspects of its development:

• harnessing its Diaspora, so that the brain drain could be turned into a "brain gain";

• moving away from a very autarchic system to become a more integrated part of the global system, which would offer significant benefits from specialization and exchange; and

• reforming the legal and regulatory regime, which had been what was holding the country back.

A Provocative Projection: China Atop the Heap

Reminding those in attendance that he had promised to provoke them, Dr. Dahlman then posted a graph projecting the next decade's GDP growth in purchasing-power parity terms for a dozen major nations (Figure 7). The projection assumes that the world's growth rate between 2005 and 2015 will be stable, and that the countries considered will grow at the rates at which they grew from 1991 to 2003. The graph shows China's economic size surpassing that of the United States by 2013 and India's surpassing Japan's by 2007. While admitting that assuming constant growth rates might be simplistic and that questions could be raised about the way purchasing-power parity is adjusted, he maintained that PPP offered "a better measure of economic size" and declared that China and India were "going to become very big players." According to a National Intelligence Committee projection of the state of the world in 2020, he added, a big factor will be rising nations—meaning China and India.

Dr. Dahlman reiterated that the key drivers of the world's increasing competitiveness are education, training, and innovation now that many of the more traditional heavy industries had faded in importance. But access to natural resources have, in contrast, remained critical; competition for energy resources would be "tremendous," and new energy technologies would be a big area to focus on. Different countries would face many different challenges, and how they respond will depend very much on their particular political, economic, historical, and sociological makeup: on where their strengths lie, how they can mobilize them, how they achieve consensus, and how they can move forward. "The countries that have been growing very fast generally have had good mechanisms for creating public-private partnerships, for consensus, and for developing a shared agenda and a vision," he observed.

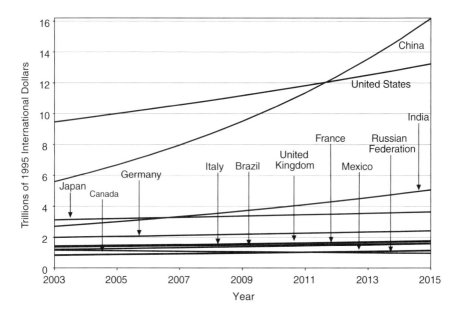

FIGURE 7 Real GDP (PPP): 2004-2015 projections using 1991-2003 average growth rates.

Concluding his presentation, Dr. Dahlman noted that China at present appears to be ahead on most measures of innovative potential, "but," he cautioned, "India is waking up."

DISCUSSION

"Just-in-Time Learning": A New Necessity?

According himself the privilege, as moderator, of opening the question period, Mr. Knox referred to Dr. Dahlman's comments on the need for "just-in-time learning" in a highly innovative economy and observed that it was a notion with which actual university teaching methods clashed. "What is going to happen in the university space to make this happen?" he asked.

Warning that his answer to this "good and tough question" would again be "provocative," Dr. Dahlman noted that universities, in general, were founded by the public sector in times of less rapid technological change. Because of the current, very rapid technological change, the private sector had had to establish many institutions of its own in order to gain access to the skills needed for rapid response. Among them had been internal institutions dedicated to training, even internal universities.

The United States has a big advantage over the rest of the world—where, for the most part, the university sector was "too isolated from the needs of the productive sector"—in having a very dynamic higher-education sector that responded very rapidly to change. But even the United States is seeing tremendous growth in online learning, because workers were in constant need of new skills and have no time to return to the campus to acquire them. While not everything can be taught online, very specialized courses could be offered. Approximately 13 percent of higher education in the United States is now taking place online, and 30-40 percent of students are no longer in the age cohorts normally associated with the universities; they were older and are obliged to come back to learn new skills of all varieties.

The United States nevertheless retains many aspects of its traditional education system, which tries to cover many different areas of knowledge. Among core skills, teaching how to learn, and learning how to learn are still very important in enabling people to go out and pursue "whatever it is that is most relevant," Dr. Dahlman said, and such concerns have "radical implications for designing education all the way from the primary through the secondary and then to the university level."

Consulting firms, he noted, have begun to discover that, rather than hire MBAs at high salaries, they can take very bright college graduates and train them quite cost-effectively to perform whatever tasks are needed. "There's going to be more and more competition from these new [education providers], some of whom will come from not the traditional university sector but even from publishing or mass media," he predicted, saying this trend is not confined to the United States but is global. The ability to use computer-based training and other forms of information technology will combine with increasing competition to put great pressure on educational systems to restructure. Although the United States, thanks to its large and exceptionally dynamic market, is very much ahead in this area, other countries are catching on.

Potential Showstoppers for China

Dr. Wessner asked Dr. Dahlman to identify the main potential showstoppers for the Chinese economy and to reflect in particular on the strength of that country's banking system and the adequacy over time of its investment capital.

Dr. Dahlman responded by outlining four key internal challenges to the continuation of what had been very impressive economic performance by China.

Environmental Concerns

On a per capita basis, China's natural resources are quite thin. The nation is very energy-dependent, a problem it has been addressing by using the very large foreign currency reserves it has amassed to acquire access to raw materials around the world. For example, China has engaged in forward-purchase programs for oil, one of which enabled the Russian government to buy back Yukos. At the same time, China has been struggling with tremendous air and water pollution, which is nearing a choke point. Despite this, however, China's leaders have opted for a very extensive expansion of the automotive sector without having found a way around the familiar environmental problems that this expansion can entail.

Economic Inequality

Inequality is growing in China both among people and regions, and it is becoming a serious concern. As television penetration is quite good throughout the country, people in the poorer parts of the west, where conditions were hard and the benefits of growth few, are able to observe the differences in living standards, something that can stimulate unrest.

The Financial System

Although the Chinese financial system appeared at first blush to be largely non-performing, it cannot be assessed by customary standards. In the absence of a system of social security, the financial system makes up much of China's social safety net. China had so many nonperforming loans because former employees of downsized state-owned enterprises, growing at a rate of 10 million to 15 million people per year, continued to receive some payment after their release. "If the country continues to grow very fast, this non-performing loan problem is not a problem," stated Dr. Dahlman, "but if it slows down, then the relative size of the non-performing loans is a big problem."

As for constraints on capital, the country's investment rate of 40 percent or more indicates that capital is not being used very effectively, something the Chinese themselves see as a "big strategic weakness." The government is setting up numerous venture-capital funds and other means of financing interesting business prospects in an attempt to address this inefficiency while also bringing in foreign banks to provide more competition, better systems of risk assessment and quality control, and better management. The desire to force this improvement is part of the reason China had joined WTO.

The Political System

Although China was moving more and more toward a market economy, it does not have a democratic political system. "At some point there is tension between people's willingness to live in a more constrained system as opposed to a freer one," Dr. Dahlman observed, saying it was not easy to predict how this issue would play out. Having seen what had happened in the Soviet Union and other countries, the Chinese authorities are extremely worried about the Internet and, in particular, the possibility it affords large numbers of unemployed from the industrial sector to organize rapidly. "They're schizophrenic about it," he said. "They see tremendous potential, but they also see the risk. And it's a very hard economy to manage."

Admitting his puzzlement at the country's success in managing its economy, and his misplaced skepticism of 20 years before that China's annual growth rates of 8 to 10 percent could be sustained, Dr. Dahlman pointed to the pragmatism of its leadership. He also suggested that, with the pie growing very fast and the benefits trickling down—sometimes to a greater, sometimes to a lesser degree—a national consensus had been achieved. "The population at large," he believed, "feels that they are getting a lot of benefits from the government, which is giving economic performance."

Possible Impact of RMB Revaluation

Al Johnson of Corning, noting that Dr. Dahlman had shown graphs using purchasing-power parity calculations, asked him to comment on the possibility of revaluation of the RMB (China's currency) and on whether a revaluation might bring forward the moment at which China and the United States reached purchasing-power parity.

Disavowing any expertise on what he regarded an important yet difficult and complicated issue, Dr. Dahlman stated that the Chinese have been very careful concerning the relative strength of their currency because they fear losing control of the speed of the country's economic growth and, as a consequence, experiencing inflation or a slowdown. While a slowdown, in the Chinese context, could mean a drop in the annual growth rate from 10 percent to 5 percent, it would still make a big difference in how benefits will be distributed and how their distribution will be perceived; the possibility was therefore considered very risky.

But because the RMB was pegged to the dollar, it had in fact been depreciating along with the dollar with respect to the yen, pound sterling, and euro. So, judging by a trade-weighted basket of currencies, the RMB has realigned considerably—it simply has not realigned with respect to U.S. currency. The United States has a particular problem in its very large trade deficit with China, with which Japan and Korea each had a trade surplus. Dr. Dahlman described, admittedly in "exaggerated terms," a relationship he saw as both difficult and symbiotic based on the purchase by the Japanese and Chinese of U.S. Treasury bonds. While the bonds paid "virtually nothing," they prevent Japan's and China's currencies from revaluing and keep the two nations very competitive with respect to U.S. markets. "They can get a hit if the U.S. devalues very rapidly because they then lose the stock value of these gigantic investments," he said, and like other investors they will "get nervous if the U.S. is subject to some big terrorist attack."

The Most Serious U.S. Flaws

Mark Myers of the Wharton Business School, who is a member of the STEP Board, asked what, from the U.S. perspective, Dr. Dahlman considered the three most serious flaws in the nation's current position.

Stressing that he was responding spontaneously, Dr. Dahlman rated the fact that only a limited number of Americans in higher education are focusing on engineering, science, and technology is a "fundamental weakness" that he said would be very hard to correct. With high pay levels in business and the legal profession attracting the nation's best and brightest, it is very important to put in place programs that provide ample grants and other assistance to graduate students in science and engineering. Enrollment in these disciplines in U.S. universities has become extremely dependent on Chinese and Indian students, and fewer are entering the country in the wake of September 11, "some because of [immigration] restrictions, some because they are now making a different choice." As a result, he said, a "big strategic weakness for the whole [U.S.] innovation system" is being exposed.

Dr. Dahlman added that articulation and coordination of policy affecting innovation needs to be increased. He acknowledged that the size and dynamism of the U.S. economy had given it a "gigantic first-mover advantage," and he also praised its flexibility and "fantastic institutional structure." Nevertheless, he called for the building of consensus, saying this does not necessarily have to be left to the government, but that there is a significant role for the private sector in self-organizing and establishing consortia where consensus building could take place. It was because the world was becoming "much tougher" and developments were taking place so quickly that this capability is in need of strengthening. Even if not persuaded by his scenarios, he said, his listeners could certainly imagine East Asia's becoming "a very big dynamo" spawning many large companies, altering the competitive landscape in the process. Europe, "sitting uncomfortably in between" East Asia and the United States, also bears watching. "All this implies the need to think very carefully about how to do strategic alliances and to have a way that the world is more balanced," he said, "so that you can have a better chance."

Finally, he pointed out that "tremendous opportunities" remained for U.S. investment in China and India. But this prospect raised the questions "national vs. multinational" and "Who is us?" This "very tricky issue," he said, "requires more thought."

Training for Innovation: From What Age?

Jim Mallos of Heliakon, returning to the theme of just-in-time learning, speculated that the character traits and habits of thought that distinguish inventors are already formed by the time a student reaches college, so that college might come too late for the training of inventors. "Shouldn't we have more emphasis on funding elementary school education when we worry about innovation?" he asked.

While disavowing expertise in the field of pedagogy, Dr. Dahlman agreed that such training should start at a very early age. One significant advantage the United States had is its reputation for "fantastic" higher education, in particular at the graduate level, which drew people from all over the world. But fundamental rethinking of education is required before critical-thinking skills can be cultivated at the very earliest level, since the country is "stuck with an obsolete structure." This is a gigantic task, and one that calls not only for continued research but also for "some real action" in the form of experimentation on what works in which setting, both within the United States and in other countries.

Remarkably, he observed, some countries have spent heavily on education yet done very poorly in the OECD's Program for International Student Assessment (PISA), an evaluation of learning skills among 15-year-olds. Finland, in contrast, has achieved very high scores on these tests with the lowest variance in performance—yet had spent far less than the average country. "How do they do it?" Dr. Dahlman asked. "I don't know. We should find out. Maybe we can learn." While PISA might be an imperfect measure of the capacity for invention or innovation, its results offered an illustration that it is not only the amount of money spent that counted, but the internal organization of an educational system that is important, including the risk and reward structure, and the incentives to be more creative. In fact, a premium needs to be placed on these features, he added.

Integrating Tacit Knowledge, Formal Research

Ken Jarboe of the Athena Alliance commented that Dr. Dahlman's coinage of "just-in-time learning" might have broken new ground, since the term described current reality far more accurately than the commonly used "lifelong learning," which was saddled with other connotations. As Dr. Jarboe understood the new concept, it is rooted in the fact that innovation is based not only on formal knowledge but also on tacit knowledge: on learning by doing, learning through experience. He asked what implications adopting the model of just-in-time learning will have for the industry-university-government research endeavor. Would it imply "some sort of just-in-time knowledge production"? And how could learning by doing or learning through experience be integrated into formal research?

Dr. Dahlman began his response with a clarification: He had not meant that very good core grounding in fundamentals could be dispensed with. Yet, he noted that many who studied the hard sciences end up working at administrative tasks where they apply very little of the deep knowledge they had acquired. That implied that a better system for producing scientists and engineers is needed, one that delivers the knowledge appropriate to the needs of the future work environment in less time. When it came to basic research, a great deal of very deep knowledge is obviously required to push back the frontier; the problem is that "too much of a narrow disciplinary approach" still reigns at a time when many different disciplines are flowing together. Therefore, mechanisms are needed that can foster and capitalize on interaction among disciplines that sometimes have no communication with one another, and that can then support work in those fields based on the possibilities and opportunities that might be found. Pilot projects are very important because of the quick response time they afforded, as are simulations. Advances in ICT enables just-in-time learning to take place in greater quantity and on a global scale; part of globalization is that it has become possible to have engineers or scientists working on a problem around the clock.

Cultural Barriers to Innovation

Jongwon Park of SRI International, saying he had recently seen data collected at the behest of the National Science Foundation on the science and technology systems of 10 Asian nations, expressed the wish to be "provocative" by talking not about the successes of Asian countries but about problems and challenges they were facing. In Korea, Japan, Taiwan, and Singapore, members of younger generations have for some time been turning away from science and engineering studies at post-secondary levels, something he called a "tremendous problem." Also, cultural barriers existing in these countries—the "Confucian, more traditional way of thinking" among them—stood in the way of spawning creativity. He asked for Dr. Dahlman's reaction.

Getting people interested in the science and technology area, where money is less easily made than in other fields, is a global problem, Dr. Dahlman concurred. While it would demand a great deal of effort, he suggested that all countries could benefit from creating financial incentives, perhaps targeted at the cost of education, along with such "moral rewards" as prizes, awards, and public prominence. Responding to the questioner's observations on the influence of Confucian thought, he attributed to cultural factors the fact that very few of the world's top-ranked universities are located in Asia. "It is something the Asian countries are going to have to face," he stated; in this case, new approaches have to be developed for application not only at the higher educational levels but far earlier as well. In effect, these problems are generic, and the present discussion is directed at conditions in the United States only because the United States was on the leading edge.

Improving Performance Evaluation

Cerise Elliott of the National Institutes of Health, noting that Dr. Dahlman had posited a need for the United States to facilitate effective communication in its information structure, asked for his suggestions for improving the evaluation and measurement of performance.

Dr. Dahlman replied that the United States has a very dynamic peer review system that works much better than those in other countries because of the United States' large size. It is very difficult to replicate this system in developing countries with very small pools of experts, where the system is subject to all kinds of biases. "When you're doing research, by definition you don't know for sure where you're going to end up," he remarked, so it is very important "to set up appropriate ways of identifying what you're targeting and how you know if you're making progress," as well as what the impact of the effort might be. He was reluctant, however, to do more than acknowledge how difficult the problem was.

Should Leadership Be a National Priority?

Shiela Ronis of the University Group, which was currently working under contract with the House Small Business Committee, asked Dr. Dahlman whether remaining the leader in science and technology should be a national priority for the United States, and whether it was important to remaining a superpower.

"I think it's hard to be a leader in everything: There are too many fields, and it's too complex," Dr. Dahlman replied, while protesting that the question went beyond his field of competence. In his opinion, the United States should focus on how to remain the leader in selected areas, and these should be not only familiar areas but also areas that, although perhaps less familiar, had potential for the future. The linkages seen among information technology, nanotechnology, and bioengineering seemed to be the beginning of a new wave, one in which many developing countries were beginning to invest quite heavily and in a very systematic way. Consideration should be given to supporting, in addition to space and military technologies, some fundamental technologies that could have big spillover effects.

New National Models

Moderators:

Bradley Knox
House Committee on Small Business
and
Charles W. Wessner
National Research Council

Briefly introducing the speakers for the day's second panel, Stefan Kuhlmann, Heikki Kotilainen, and Peter Nicholson, Mr. Knox thanked all three for traveling considerable distances and said those in attendance were anticipating their contributions with interest and excitement. He called on Dr. Kuhlman to present first.

THE RECORD AND THE CHALLENGE IN GERMANY

Stefan Kuhlmann
Fraunhofer ISI, Germany

Dr. Kuhlmann expressed gratitude at being offered a chance to participate in what he characterized as an interesting and challenging event being held in an environment that was impressive both institutionally and architecturally. He said that he had been asked to speak about the status of research in Germany and the country's research innovation system, whose development is increasingly becoming part of a European research system.

Dr. Kuhlmann then outlined a presentation that was structured in four parts. Part one laid out the strengths and weaknesses of the German innovation system. Part two briefly introduced the governance structure of innovative research in Germany and the related institutional landscape. Part three delved into innovation policy and programs, at both the level of the German federal government and the level of the Länder, which in the German context was the equivalent of the U.S. state level. Part four considered a number of current developments and challenges for the future.

Strengths of Germany's Innovation System

Germany's strength is that it has been and continues to be highly "innovation oriented." Its gross R&D expenditures have been running at just under 55 billion euros, or around 2.5 percent of GDP. Quite strong in the 1980s, the country's spending on R&D had fallen during the early 1990s in conjunction with reunification but has been increasing in recent years. Companies account for 66 percent of R&D expenditure, a considerable share in Dr. Kuhlmann's view, and Germany leads the European Union (EU) in the percentage of small and medium-sized enterprises (SMEs) innovating in-house. Germany's 14.9 percent of the world market for R&D-intensive goods places it second behind the United States, and it is in the EU's top three in share of manufacturing sales attributed to new products. Its number of patent applications per inhabitant, 127, is second-highest among large countries, and it ranks third among all nations in international publications with 9 percent of the total.

Dr. Kuhlmann noted that in 2001, business provided the largest share of R&D financing by far, followed by the federal and Länder governments. A small amount of R&D expenditure came from abroad, from either the European Commission or private-sector sources, which had been funding R&D to be performed in Germany with growing frequency. Business dominated performance of R&D as well, followed by the higher-education sector and by nonuniversity research institutions, the latter being a feature typical of the German system and one that he promised to return to later in his talk.

Germany's Weaknesses

Calling the persistence of risk-averse behavior among banks one of the major weaknesses of the German research system, Dr. Kuhlman noted that financing innovation has become increasingly difficult, especially for SMEs. He then listed some of Germany's other weaknesses:

- It is suffering a clear loss of momentum in some, although not all, of the high-tech sectors in which it had been strong—among them, pharmaceuticals, computers, electronics, and aircraft.

- Its technological performance has come to depend increasingly on the automotive sector.

- East Germany and Berlin together eat up nearly one-quarter of the 9 billion euro federal research budget while employing only 11 percent of the country's R&D personnel and accounting for only 6 percent of its patent production. Yet, East Germany is "not very strong in competitive and innovation terms" despite the amount of money invested there.

- The performance of Germany's educational system, long considered quite strong, has declined to the point that Dr. Kuhlmann talked of a "crisis" and envisaged an expensive restructuring.

- The growth of the public sector's R&D spending is lagging that of the private sector over two decades ending in 2001. More recent data would show a slight increase in public and some stagnation in business R&D investment, he noted.

Governance Structure of the German Innovation System

Turning to the complex governance structure of Germany's system of innovative research, Dr. Kuhlmann listed three levels in descending order—Federal (National), Länder (State), and Regional (substate groupings)—that he said were overarched by an "increasingly relevant" Supranational level, represented specifically by the European Union.

Existing on the national level in Germany, exemplifying a phenomenon found in many European countries, is "a kind of competition" between the Federal Ministry of Research and Education and the Federal Ministry of Economics and Labor. Although there is collaboration within the government, there is some duplication as well, and it is not clear who was in charge of research, technological development, and innovation. At the state level, interstate competition mirroring that found in the United States is on the rise. Relations between the federal government and the Länder are marked by "coopetition," with initiatives at the two levels sometimes in conflict, sometimes complementary. The regional level depends on the federal and state levels for investment.

While governance at the European level is gaining in importance, the role of EU investments is of greater relative significance in the smaller nations among the EU-25. Germany, Dr. Kuhlmann estimated, receives some 5 percent of the R&D funding that comes through EU channels. Still, there are exceptions according to sector: In information and communications technology (ICT) in Germany, where national programs had not been very strong, EU funding has played a more important role than in other areas. Despite much discussion of "subsidiarity"—as complementarity among initiatives at the EU, national, and state or Länder levels was called in official EU parlance—there is, in practice, a great deal of both overlap and confusion.

The Institutional Landscape of German Research

Dr. Kuhlmann noted that the German research system's institutional landscape varies from basic to applied research, with funding sources spanning from public to private organizations. While industry, which contributes the lion's share of the country's R&D investment, is mainly occupied with applied research, it also participates in some areas of fundamental research. Universities receive state and federal funding, though they increasingly receive money from industry and conducted applied as well as basic research. The 8 billion euros received by universities covers research as well as teaching.

Dr. Kuhlmann then focused specifically on the nonuniversity research institutions: the Max Planck Society (MPG), which does excellent basic research; the Helmholz-Gemeinschaft (HGF), a national organization with 15 major centers doing research in many fields and particularly in "problem-driven" areas; and his own organization, the Fraunhofer Gesellschaft (FhG), which does applied research on a contract basis and receives a small amount of institutional funding. Problems beset collaboration among the actors occupying the contract-research area, he said. But collaboration with industry is "quite well developed" and, as such, no longer the issue it was in the 1980s and 1990s.

He next turned to the basic mechanisms of federal R&D funding in Germany. The leading category, at nearly 47 percent of the total, is "institutional funding," which goes to such nonuniversity research organizations as Max Planck, Helmholz, and Fraunhofer. Project funding, routed through programs at the federal level, represents 40 percent; its share, greater in the 1980s, decreased through the 1990s as institutional funding rose. The role of project and program funding in innovation policy is thus exhibiting some shrinkage in Germany, although not to the point that it has lost its relevance.

Innovation Policy Programs: BMBF

The two main providers of funds at the national level, as mentioned above, are the Federal Ministry of Education and Research, or BMBF (Bundesministerium für Bildung und Forschung), and the Federal Ministry of Economics and Labor, or BMWA (Bundesministerium für Wirtschaft und Arbeit). The BMBF supports a broad variety of technology and innovation programs—so broad, in fact, as to be difficult to track.

For this reason, said Dr. Kuhlmann, presenting a full list of these thematic programs is infeasible. Although there are official lists, none of these lists reflects a precise technology or innovation program. He and coworkers therefore assembled a table (Figure 8), that encompasses 3.4 billion euros' worth of programs. These range from biotechnology and nanotechnology programs to initiatives on new materials and electronics. Some fund technological research, but many contain an element of innovation support. This element, when present, owes its

Program	Year	Funding 2003 (Euro)	Tech. Focus	Coop. / Networks	Regional focus	East Germany
Thematic Programs, e.g., biotechnology (109 m), nano/new materials (89 m), electronics (81 m)	Since 1970s	3.4 bn	✓	✓		
InnoRegio ("Enterprise Region")	1999-2006	65 m		✓	✓	✓
Innovative regional growth poles (Wachstumskerne)	2001-2008	24 m		✓	✓	✓
Applied R&D at universities of applied sciences (FH3)	1992-2008	12.5 m		✓		
EXIST—Start-ups from science	1998-2005	8 m		✓	✓	

⇨ Furthermore: support of competence networks and for patenting activities, etc.

FIGURE 8 Innovation policy programs, federal level: BMBF (selection).
SOURCE: Adapted from Federal Budget BMBF 2003, BMBF 2004.

existence in large part to the programs' having been conceptualized as cooperative; that is, as a basic precondition for getting them funded, they have to include
some degree of collaboration between researchers from the public sector, including academia, and industry.

The table's remaining categories are likewise incomplete, Dr. Kuhlmann
said, and for similar reasons. Among them was the InnoRegio program, which is
aimed at networking in regions of eastern Germany. Also listed were: "Innovative
Regional Growth Poles," a program similar to InnoRegio; support for research
in the Fachhochschulen (FH3), or universities of applied sciences; and EXIST,
which helps universities develop a startup-friendly climate. He pointed out that
a significant feature of this admittedly incomplete table is that funding for the
four programs aimed specifically at innovation is small compared to funding for
the thematic programs.

Innovation Policy Programs: BMWA

Next came a comparable table showing innovation-policy programs run by
the BMWA (Figure 9), which, as mentioned earlier, is to some extent in competition with the BMBF. Heading the list is a program focused on innovation in
East Germany, followed by the largest program in monetary terms, FoKo/Pro
Inno, which targets collaborative work in innovation by SMEs. The next two
programs were of similar size: The Promotion of Joint Industrial Research, a
bottom-up research funding mechanism driven by industry and cofunded by the

Program	Year	Funding 2003 (Euro)	Type	Coop. / Networks	Regional focus	East Germany
Special R&D Program for East Germany / INNO-WATT	1990–2003 / since 2004	104 m	Subs.			✓
FoKo / Pro Inno / Pro Inno II	1993-1998 / 1999-2003 / 2004-2008	158 m	Subs.	✓		
IGF - Promotion of Joint Industrial Research	Since 1954	97 m	Subs.	✓		
European Recovery Innovation Program	Since 1996	95 m	Loans			
InnoNet	1999-2005	13 m	Subs.	✓		
NEMO	2000-2004	6 m	Subs.	✓	✓	✓

⇨ Furthermore: a variety of loan and venture capital programs

FIGURE 9 Innovation policy programs, federal level: BMWA (selection). SOURCE: Adapted from Federal Budget BMBF 2003, BMBF 2004.

ministry; and the European Recovery Innovation Program, which provides loans to innovating companies. Rounding out the table—which, like that for BMBA, is incomplete—are a pair of very small programs, InnoNet and NEMO, the latter an acronym standing for Network Management East Germany. Critics deride this profusion of programs in Germany as an "innovation-policy funding jungle" and claim that no one can understand it fully, a charge both ministries rejected. While acknowledging that, as a political scientist, he had his own ideas about the origin and development of this funding structure, Dr. Kuhlmann said that on the present occasion he would keep them to himself.

Programs Run by the Länder

Conceding that it would only deepen the complexity of the policy picture, he then turned to the Länder, which are responsible for funding and operating the nation's universities, the latter traditionally receiving most of the research money available to the former for distribution. But the Länder are increasingly going beyond supporting university research to set up programs designed to:

- spur technology development by funding both single-organization and cooperative R&D;
- foster technology transfer, including transfer of personnel;
- support startup companies by providing consulting, coaching, incubation, and financial assistance;

- furnish venture capital and loan guarantees;
- encourage patenting activities; and
- participate in technology parks and incubators.

Huge differences in investment existed from state to state. But while, for example, Bavaria accounts for just over one-fifth of total nonuniversity research spending by the Länder and Saarland for a mere 0.5 percent, EU and other sources supplement funding in weaker regions.

Program Anatomy: Pro Inno

Dr. Kuhlmann chose two examples from this broad variety of programs to examine in detail. The BMWA's Pro Inno, which had been running for more than 10 years and through which 630 million euros were invested in the period 1999-2003, has as its goal increasing the R&D capability and competences of SMEs. This is to be achieved through collaboration on not only a national but also an international level. By this he meant that German companies could receive money in support of efforts involving partners outside the country. Pointing this out as "an interesting feature" of the program, he said that "internationalization is now understood as a relevant part of future R&D policies." Subsidies under Pro Inno run at between 25 and 50 percent of the cost of R&D personnel ranging across four program lines—cooperation with firms, cooperation with research organizations, R&D contracts, and personnel exchange—with multiple applications totaling up to 350,000 euros per firm allowed.

Since 1999, 4,850 firms and 240 research organizations have participated, with 4,000 R&D employees per year engaged in Pro Inno projects. A past evaluation in 2002 had shown that some three-fourths of participating firms would not have conducted the R&D had it not been for the program; Dr. Kuhlmann and colleagues were preparing an updated evaluation due later in 2005.

Considered one of Germany's most significant R&D programs, Pro Inno is widely known and accepted among its target group. Its outstanding characteristics are its broad, open approach, high transparency, easy access, and relative lack of bureaucracy. The program receives a high number of applications—even though there is no guarantee of funding due to budget constraints and despite significant problems, during market entry.

Program Anatomy: InnoRegio

The second example, BMBF's InnoRegio program, is aimed at strengthening the endogenous innovation potential of weak regions in eastern Germany by setting up sustainable innovation networks. A complex, "multiactor" program, it encompasses not only SMEs, large companies, and research organizations but also many other public and private activities and initiatives, funding both network management and projects aimed at developing products and services. Run as a competition, the program has three stages: a qualification round in which ideas are solicited for potential networking initiatives (444 proposals were submitted in 1999); a development round in which the regions selected (25 of the 444) refine their concepts and projects; and a realization round in which the winners (23 of the 25) receive multiyear financial support for their initiatives.

The InnoRegio program is linked to an increase in innovation activities. In the prior two years, two-fifths of the firms selected by the program have received patents and almost all have introduced new products. Since 2000, 50 new firms have also been founded. But this last fact is less impressive in Dr. Kuhlmann's judgment than, what he considered the program's main impact—i.e., the creation of innovation networks across eastern Germany that comprises both public and private actors. Nevertheless, he noted that "in East Germany there is still a tendency to expect public funding to be the main source of stimulus towards innovation." Moreover, although there is scant disagreement that networking is a basic precondition for innovation, the question of whether the impact of such initiatives will be sufficiently long-term, stable, and robust to justify the investment they require remains unanswered.

Germany's "Partnership for Innovation"

Taking up a theme that Dr. Dahlman had earlier discussed in the U.S. context, Dr. Kuhlmann suggested that, in light of the variety of public policies affecting innovation, major countries may see the need to introduce some kind of coordination and collaboration across the agencies responsible for such policies. Germany's current response, a federal innovation initiative, was born in October 2003, when Chancellor Schroeder summoned representatives of various stakeholders in the field—policy actors, public agencies, major companies, associations of SMEs, and major research organizations—to a forum where they debated challenges for the country's innovation performance. In the aftermath, they created the Partnership for Innovation and started some longer-term "pioneering" activities. The overarching aim of this initiative's broad agenda is to improve the framework for innovation in Germany through the collaboration of public and private actors.

The Partnership for Innovation had established a number of working groups of stakeholders, some of which Dr. Kuhlmann and his colleagues contribute to. "In some areas there is just talk, and they will continue just to talk," he said. But "in some areas these working groups have actually started policy initiatives, regulatory [and institutional] reforms, and so on."

Under the rubric of "High-tech Masterplan," an effort is under way to ease access to venture capital; while some activities it comprised had predated it, a high-tech startup fund was kicked off only weeks before the symposium with 10 million euros, and is expected to grow in time. As the initiative grouped such a wide variety of activities, its efficacy will not be easy to evaluate, he said.

Germany's Future: Threats and Opportunities

To conclude, Dr. Kuhlman provided an appraisal of the threats and opportunities Germany was facing, as well as a rundown of the federal government's immediate priorities. He began his enumeration of the threats by emphasizing Germany's high degree of dependence on its automotive cluster. Even if it had not hurt the country yet, this dependence is leaving Germany open to challenge in the longer term, particularly from East Asian competition. Second, he cited an anticipated shortage in the supply of highly qualified labor, especially in the engineering field. Like their contemporaries in the United States and East Asia, young people in Germany are not studying engineering and science in sufficient numbers. But Germany is also laboring under the problem of not having a big enough population of high-skill workers and of having had, until recently, regulations restricting immigration of engineers and scientists. Third, he reiterated that the country's technological performance is losing momentum in some areas. Finally, he noted that public financing for the national science infrastructure is shrinking relative to financing coming from the private sector.

Nevertheless, Dr. Kuhlmann said, there were opportunities. The technological and marketing prowess of Germany's automotive sector might allow it to turn the challenge it was facing to its benefit. Efforts were being made to build on the strong position that Germany, like other traditional EU members, enjoy in the growing markets of both Eastern Europe and East Asia. The country has an excellent science base in such advanced fields as biotechnology, chemicals, and nanotechnology. And its labor force remains highly qualified, at least in comparative terms.

The Potential of European Integration

Finally, Dr. Kuhlmann expressed his personal conviction that Europe's integration, not in the political arena but in the area of research systems, could lend Germany's national system huge potential. Possessed of the biggest research system in Europe, Germany has in the past tended toward both a domestic orientation and a degree of complacency. Increasingly, however, major research organizations and individual researchers alike are learning that they needed not only to collaborate on the European level, but also to reestablish themselves institutionally in a European context, if they wished to avoid overlap and to create new clusters of strength with worldwide visibility and potential. These developments were so recent that, he cautioned, some of the impressions he was conveying were still in the realm of speculation. While the single market offered significant opportunity, national governments, and particularly the German government, were yet to comprehend the existence of this potential and the difference between national and European. While this had been the source of some difficulties for those in charge, there was, he asserted, "no alternative to more collaboration and integration on this level."

Identifying the immediate priorities for the German federal government, Dr. Kuhlmann said there is recognition of a need to improve linkages within the country's fragmented research system, whose existing boundaries are rigid to the point that horizontal professional mobility was scarcely possible. Changes should, he recommended, be designed to take account of European integration as well. Meanwhile, work needed to reform the country's university system with the goal of achieving international excellence has begun, although in his own opinion, progress has been slow compared to that observed in other countries, both within and outside Europe.

THE TEKES EXPERIENCE AND NEW INITIATIVES

Heikki Kotilainen
Tekes, Finland

Dr. Kotilainen expressed his appreciation for the opportunity to present Finland's "small-country approach to innovation policy." His presentation began with a very brief statistical overview, and continued with descriptions of Finland's innovation system. He concluded with examples of the system's achievements.

Before embarking on it, however, he would acknowledge a handful of factors that could be seen as drivers of change and, at the same time, challenges to be met. Various phenomena surrounding globalization discussed earlier are having quite an impact on Finland, as many companies move manufacturing and other operations to such countries as China, Malaysia, and Indonesia. Finland is also undergoing a very serious demographic change of the kind affecting all of Europe, a problem for which solutions are not yet become apparent. Sustainable development is another challenge commonly faced in the industrialized world, as is the management of knowledge and competence; in the latter, Finland, as a small country with limited resources, has to exercise great prudence.

Technology and networking are two drivers, or challenges, that Dr. Kotilainen viewed as one: When companies move, countries find themselves having to find new technologies to replace those that parted, while the companies, because they applied R&D independently of where it is performed, need to be on the lookout for technological and business alliances. Finally, he noted, change in the character and dynamics of innovation constitutes a challenge in itself.

Dr. Kotilainen noted that Finland is placed near the top of the world rankings in science and technology, second only to the United States. How could a small country whose only true natural resource are its people accomplish this? What had happened, he said, was that Finland's industrial structure changed tremendously over the previous four decades. The pulp-and-paper sector, which accounted for a little more than two-thirds of the country's exports in the 1960s, saw its share fall to below one-quarter by 2004, while exports in the electronics sector, including telecommunications, had taken off.

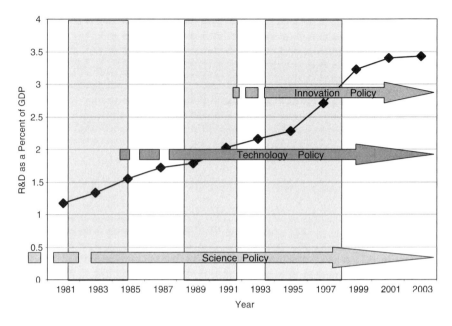

FIGURE 10 R&D/GDP in Finland, 1981-2003.

A major factor in this change, according to Dr. Kotilainen, has been "conscious and continuous investment" by Finland that has raised its R&D spending as a percentage of GDP from around 1.5 percent in 1985 to nearly 3.5 percent by the beginning of the current decade. Coinciding with this increase has been the evolution of Finnish policy in the realm of science and technology (Figure 10). Even though Finland's science policies have been in place since the 1950s, technology policy formulation did not happen until the mid-1980s, following the establishment of Finland's Science and Technology Policy Council. It was only after it began to implement technology policy that innovation policy appeared on the agenda. While Dr. Kotilainen attributed this progression to chance rather than to any special wisdom—and noted that, in fact, a similar evolution can be seen in other countries—he said that it had turned out to be a fortunate one. For innovation policy, had it made its appearance prematurely, might have been met by silence from the government and suffered a consequent loss of credibility. "You cannot," he remarked, "jump from pure science to innovation immediately."

Dr. Kotilainen pointed out that private-sector spending has shown the most growth since 1985 and has accounted for some 70 percent of the current total investment, which has been running at around five billion euros annually. Even so, the public sector has been the prime mover: Only when the government began to take R&D seriously and increase funding for it did private investment follow. Still, when it comes to public funding for R&D taking place in companies, Finland is well below the Organisation for Economic and Co-operative Development (OECD) average at less than 5 percent of total corporate R&D spending. The position of some of the more highly ranked countries is, however, the result of heavy investment in military research, which Finland does not fund. The sector of the Finnish economy most active in R&D is electronics and telecommunications. Nokia alone accounts for something like 40 percent of Finland's private sector, and while a small country can count itself lucky to have such a "behemoth," the firm's relative size could also be seen as a "threatening issue," Dr. Kotilainen said.

Finnish Innovation System's Institutional Structure

Turning from a quantitative to a qualitative portrait of innovation in Finland, Dr. Kotilainen identified the two main public organizations in the domain of R&D: the Academy of Finland, operating under the country's Ministry of Education and charged with basic research; and his own institution, Tekes, operating under the Ministry of Trade and Industry and charged with applied research. Public-sector R&D actors also included universities; VTT, a large and multidisciplinary research institute under the Ministry of Trade of Industry; and other ministries, such as Agriculture and Forestry, that had their own research facilities. Providing direction to government institutions through achieving consensus on the growth of R&D is the Science and Technology Policy Council, which dates from 1986 and is, he said, "a very important element" in the Finnish system.

Understanding how the game of innovation was played in Finland demands an acquaintance with the Science and Technology Policy Council. Chaired by the prime minister, the council has as members five other cabinet ministers, including the minister of Finance; the directors general of the Academy of Finland, Tekes, and the VTT; and representatives of the universities, industry, and the labor unions. Every three years, the council issues an outline for science and technology; this outline takes the form of recommendations rather than regulations, something Dr. Kotilainen saw as strengthening Finland in comparison with the many countries where the government assumed an operational role. On the basis of the outline, institutions like Tekes work with the ministries to prepare annual objectives, which in turn become the basis for an agreement covering what is to be done over the coming year. The implementing institutions, after executing the plan, report to the ministries, which then report to the Council. Showing a schematic diagram of the process (Figure 11), Dr. Kotilainen pointed out that the spheres of planning and implementation were separated by a dotted line. "The operative part is operative and the planning is planning," he emphasized. "They are not mixed."

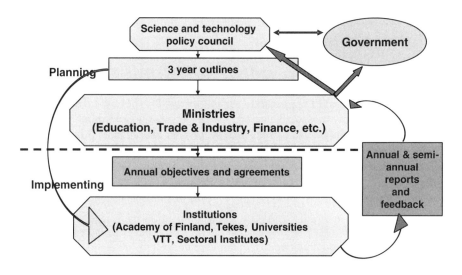

FIGURE 11 Planning and implementation of technology and innovation policy.

Implementation is left to the "expert organizations," as he qualified Tekes, which might receive funding from the government but are free to act as they see fit. He then displayed a table providing details of an increase in funding for innovation proposed in 2002 for the five years that were to follow (Figure 12). While the Ministry of Finance was free to accept or reject such recommendations, it generally went along with those receiving strong support—and, as the table shows, the funding increase for innovation was already well under way.

For Finns R&D Is Investment, Not Expenditure

Having dealt with the system's structure, Dr. Kotilainen took up the subject of its operation. In Finland, R&D funding is considered to be an investment rather than an expenditure. Money that goes directly from funding agencies to companies is seen as short-term investment; money that flows through universities and research institutes, creating technology and new skills that then contribute to companies' competitiveness, is seen as long-term investment. The former route might require a few years at most, the latter a decade or more, but both types of investment are considered necessary. And not only that, they are considered one of the best investments the government could make, since they ultimately bring back sums 10 to 20 times their original size in the form of taxation.

	Research funding, million euros			Increase in other funding, million euros
	2002	increase	2007	2007
Universities				
- research funding	375	45	420	
- other increase in basic funding			105	
Funding organisations				
- Academy of Finland	185	70	255	
- National Technology Agency	400	120	520	
Other research funding				
- research institutes	235	40	275	
- ministries	205	25	230	
Total				
- research funding	1400	300	1700	
- other funding				105
Increase, total	300			+ 105 = 405

FIGURE 12 Recommendations of the Science and Technology Policy Council of Finland relating to research and innovation funding.
SOURCE: Tekes.

Tekes's Funding Strategy, Operations

Tekes itself has a steadily if gradually rising budget that reached 430 million euros in 2005. Research funding in the form of grants and company funding in the form of both grants and loans is distributed through a variety of instruments, such as national technology programs, direct company R&D funding, direct research funding of universities and research institutes, as well as equity funding for start-up entrepreneurs. Acknowledging the instruments themselves to be "very standard," he placed emphasis on Tekes's implementation. As networking and cooperation takes place between the companies and universities involved from the very beginning, all the instruments could be regarded as technology-transfer instruments. In fact, with the exception of TUPAS, a "little" program designed exclusively for SMEs, Tekes has no separate technology-transfer procedures. Private-sector entities are always required to provide matching funds when participating in university research, while companies receive credit if they invited universities to join research programs they had initiated.

In 2004, Tekes invested around 409 million euros, with 237 million euros going to companies—of which 31 million was in the form of reimbursable, reduced-rate R&D loans, 41 million was in capital loans, and 165 million was in R&D grants—and 172 million euros going to research institutions in the form of grants. In that year, about half of all funds were channeled through established technology programs while the other half went to individual, "bottom-up" projects that were presented as unsolicited proposals by industry or research institutions.

Such applications are accepted at any time from either sector, but Tekes always considers whether they contained a component of cooperation between the two sectors. The established programs, seen as highly important to Finland's general technology development, fund large projects featuring cooperation among multiple companies and universities or nonuniversity research institutions. The proportion of funding going through each of the two channels is not mandated by the government. In fact, the allocations vary year to year at Tekes's discretion and depending on customer needs.

Placing a Premium on Cooperation

Although Tekes provides support on different terms under its technology programs (according to whether it was destined for public- or private-sector entities) this fact does not prevent projects from cooperating across sector lines. In fact, Tekes sees implementation of results from public research as dependent on parallel execution and networking with company projects, even to the point of pooling personnel.

In 2004, Tekes funded projects under 23 existing technology programs covering a wide range of emerging technologies having an overall value of 1.2 billion euros; such programs typically were of 3 to 5 years long and, in any given year, drew 800 separate "participations" from public research units and 2,000 from companies.[9]

That Tekes provides no more than 60 to 80 percent of the funding for university projects indicates that it always requires matching funds from industry, Dr. Kotilainen explained, adding that it generally refrains from funding projects at 100 percent even though it has the ability to do so. EU state-aid regulations limit support for private-sector projects to between 25 and 50 percent. Tekes's decision-making and coordinating functions are guided by a steering committee chaired by a representative of industry. "We don't let the professors be chairman," he commented.

To carry out its functions, Tekes needs a facilitator in the form of national programs to link supply and demand, where supply is considered to be research and demand to be the enterprises. It is Finnish companies' ability to articulate their needs, not just for the next year but over five years, that enables Tekes to design effective programs. "With our small resources we cannot do research for [the sake of] research," Dr. Kotilainen explained. "It should be relevant to our economy." Similarly, the level of the research funded is expected to be appropriate: "We cannot target a Nobel Prize in each field—that's not for us." Appropriate

---

[9]For a list of current technology programs, see <http://www.tekes.fi/english/programmes/all/all. html#Ongoing>.

goals, furthermore, help ensure that companies have the technology appropriate to their purposes and have the capability to absorb new technologies for their use. He named "this linkage, through the national programs, between research and the companies" as one of the reasons that Finland places "fairly high in the competitiveness statistics."

Unique Features of Finland's Innovation System

While it might depend on "normal" funding instruments, the Finnish innovation system boasts some unique features. First among these is *trust* among participants in the the university-government-industry "triple helix," that by the early 1980s, enabled genuine and voluntary cooperation. *Lack of corruption* is a second advantage, although in Finland's context "corruption" is likely too strong a word, said Dr. Kotilainen. "We don't have to expend energy to avoid fraud in the system," was how he rephrased it. A third factor is *consensus* and the Science and Technology Policy Council helps ensure a high degree of agreement, which in turn enhances implementation. A fourth factor is *cooperation*. Tekes and the Academy of Finland work very closely together, helping make possible simultaneous funding of universities and companies, and thereby the coupling of research with development. Such efforts extend beyond science to the coordination of technological development with social development, a "very important" consideration, he said, as the country "cannot leave part of the population to be dropouts [when it comes to] technological development."

The small number of actors in Finland's system, with its consequent simplicity and "holistic" quality, constitute another advantage. Tekes itself enjoys a very high degree of independence from the cabinet and the ministries in assessing projects and making decisions on funding, both of which it is free to do independently in-house. Its experts working on assessments have access to industry and research leaders, and, like all the institution's employees, possess knowledge of both the domestic and international markets suited to their responsibility over international cooperative efforts.

Impact of Tekes's Activities

Besides the increase in R&D investment, the system has helped generate a remarkably sharp increase in high-tech exports' share of Finland's total exports, from below 5 percent in 1988 to over 20 percent beginning in 1998. To be seen specifically among corporate R&D projects funded by Tekes is a prevalence of cooperation. Networking—whether with SMEs, foreign corporations, or research institutes and universities—is occurring in around 80 percent of all efforts, as a result of which Finland leads the EU in intercompany cooperation.

Coinciding with Tekes's investment of 409 million euros in 2004, Dr. Kotilainen stated, 770 new products have reached the market and 190 manufacturing processes have been introduced. The institution also claims to have contributed to 720 patent applications, 2,500 publications, and 1,000 academic degrees at the bachelors, masters, and Ph.D. levels in that year. According to a 1999 study of public financing's effect on companies' innovation activities, conducted by VTT, the receipt of funding from Tekes often causes project goals to be reset higher than they had been initially. This study also concluded that Tekes funding allows for the expansion or acceleration of projects, the latter being of particular importance in sectors where the first company to market takes all. A study by Finland's National Audit Office dating to 2000 found that Tekes funding allowed not only broader and more rapid implementation of projects, but also surmounting of risk barriers: 57 percent of projects considered in the study would simply not have been undertaken without support from Tekes. Finally, a pair of 2003 studies by the Research Institute of the Finnish Economy (ETLA) found increased corporate R&D investment to result from public investment in research and not the other way around.

Innovation: More than Research

Returning to the challenges accompanying globalization and other changes in the nature of innovation, Dr. Kotilainen observed that, while research is important for radical or disruptive innovations, most innovation is incremental and not necessarily rooted in research. When it comes to user-based innovations, the customer is paramount. The tendency for manufacturing to merge with services is an important development. Smaller firms are more directly engaged in innovation itself, while large companies, he implied, are more involved at the level of application. The growth in importance of multidisciplinary research underlines the value of alliances and other forms of cooperation. For these reasons, the public and private sectors need to innovate together.

Dr. Kotilainen concluded with a brief assessment of the implications of these challenges for his own institution. The role of the innovation agency is changing along with the landscape, a development that requires serious thought about the future. No longer regulators, agencies like his have to become instructors, he said. "We are not a referee [but] should be a partner in this process; we are not a system integrator but a networker; [we are] not the finance provider but [an] investor." This, in turn, demands a big change in the mindset of the agency, which should no longer be "only an administrator" but "should be innovative as well if we ask innovations from our customers."

CONVERTING RESEARCH TO INNOVATION

Peter J. Nicholson
Office of the Prime Minister, Canada

Dr. Nicholson expressed his desire to clarify at the outset that, as chief of policy in the office of Canada's prime minister, he is not specifically charged with management of science and technology. Although experienced in the field as a former faculty member at the University of Minnesota, he had never administered science and technology programs in government. Much of what he was about to say, therefore, was distilled from large quantities of good advice he had received from people dealing directly in such matters.

As he judged it unnecessary to explain why Canada should be concerned about science-and-technology and innovation policy, Dr. Nicholson immediately began crafting a statistical portrait of the country's R&D efforts. In this, he explained, he would be aided by graphics showing amounts in U.S. dollars that had been converted from Canadian dollars at the purchasing-power parity of roughly 85 U.S. cents to a Canadian dollar. On a chart tracking Canada's gross domestic expenditure on R&D as a percentage of GDP in the 10 years from 1993, he pointed out a steady upswing in spending that peaked with the end of the tech bubble in 2001. The ensuing downturn could be attributed to a fall-off in R&D spending by business, particularly in the information-technology sector and more particularly by Nortel.

Higher Education Funding "Remarkably High"

Canada's current total R&D spending is around $19 billion a year. Business enterprise expenditure on R&D accounts for around 55 percent of the country's total, government intramural expenditure on R&D for around 12 percent; the remainder comes from higher education expenditure on R&D, which, at a "remarkably high" 33 percent, places Canada atop the G-7 and third among OECD nations. "Otherwise we're down in the middle of the pack," he said, "notwithstanding the increase [in Canada's R&D spending] over the years. The point is that the bar has been rising."

Even if Canada does not post world-leading ratios of R&D expenditure to GDP, its overall economic performance suffices to make it the second wealthiest among mid-sized countries in the OECD based on GDP per capita and second in the G-7 as well. Much of the technical dynamism in the Canadian economy is due to its unique level of integration with the U.S. economy. "In many, many sectors, there is one economy," said Dr. Nicholson, adding that "a great deal of technical sophistication in the Canadian economy is embodied in imported capital." To illustrate, he noted that a North American car buyer is unlikely to know whether the Chevrolet, the Chrysler, or even the Toyota he or she was buying is manufactured in Canada or the United States.

Fiscal Consolidation Opens Gate to Innovation

Contributing to Canada's positive performance is a fiscal consolidation that has taken place over the previous 13 years. Canada's fiscal problems had been rapidly growing out of control, but in 1995—for reasons that, according to Dr. Nicholson, remained "a little mysterious"—Canada "got an incredible dose of fiscal religion, which has continued to this very day." The federal budget went into the black in 1997 and has stayed there, unlike in the other G-7 countries, which have all gone back into deficit. This turnaround has permitted the Canadian government to make numerous choices that up to that point it has not been able to make, chief among which is a determination to fortify the microeconomic foundations of the economy. Traditionally, Canada's economy has been resource based, with a great deal of industrial power added through its integration with the United States. "If we wanted to have something that was home-grown and that could give us a degree of independence," he explained, "we had to build our innovation capacity from the ground up." It was this effort that would be treated in the rest of the presentation.

Dr. Nicholson projected a graph tracing the evolution of federal spending on R&D in Canada (Figure 13), which demonstrates the "paradigm shift" in federal support for higher education that coincided with the federal budget's going into surplus. Federal support for R&D in the business sector has trended down in real terms over the same period (somewhat more than indicated in Figure 13, which takes no account of inflation.) Although support for intramural R&D within the federal system has risen, research has been overwhelmingly outsourced to higher education, primarily to the university and the research hospital systems.

A Policy Conundrum: Private vs. Public Return

To help explain the public-policy rationale for supporting different kinds of activities related to innovation, Dr. Nicholson displayed a chart representing a spectrum running from basic research to such unquestionably commercial activity as marketing (Figure 14). The relative rate of social return declines as one moved rightward from the left-hand end of this chart. There is, nonetheless, some private return at the left-hand end, and perhaps even a great deal in the case of ideas protected by patents. Conversely, some social return occurs at the right-hand end of the chart owing to spillovers. But what he described as a conundrum for innovation policy focused on a "potentially 'orphaned' domain" near the middle where it is difficult to discern which of the two types of return is higher; the ambiguity of their relation to one another has occasioned such policy arguments as the one, alluded to earlier, concerning Sematech.

Canada's federal government is active all along this continuum through many dozens of programs, some of them quite large. These include the Canada Research Chairs (CRCs), the Canada Foundation for Innovation (CFI), Technology Partnerships Canada (TPC), and the Industrial Research Assistance Program

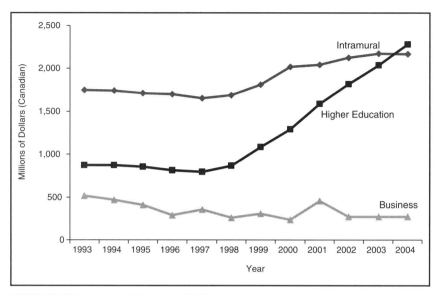

FIGURE 13 Evolution of Canada's R&D.

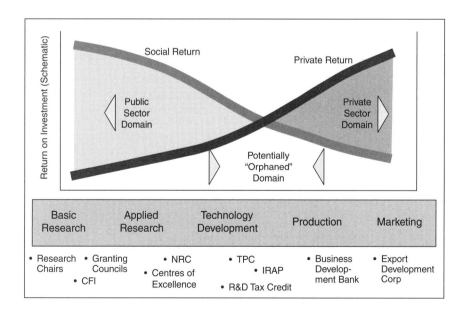

FIGURE 14 Canada's support programs span the innovation continuum.

	CREATED	MISSION	FED FUNDING (millions of U.S. Dollars)*
Granting Councils	Various dates	Support university researchers in Sciences, Engineering & Medicine	$1,660 Annual**
Foundation for Innovation	1997	Create top-flight research infrastructure in Universities & Hospitals	$3,100 Invested
Research Chairs	2000	Support top scholars to build world-class research capacity in universities	$765 Invested
Tech. Partnerships	1996	Provide targeted assistance to high-risk R&D commercialization	$255 Annual
IRAP	1962	Build innovation capacity of SMEs thru technical advice & assistance	$135 Annual
BDC Early-Stage Venture Capital	2004	Seed investment to leverage private VC in new technologies	$210 Invested

FIGURE 15 Federal support for innovation: Examples.
NOTE: Federal funding is labeled annual in some cases and in others covers the investment for the full period. (*)—Purchasing-Power Parity estimated at USD 0.85 per CDN dollar. (**)—Includes indirect costs of research and graduate student support.

(IRAP). But before turning to describe them, Dr. Nicholson estimated that the R&D tax credit generates a benefit of $1.5 billion to Canadian corporations conducting research. In cases of companies that are not publicly traded, he specified, these credits are refundable, so that benefits are not available only to firms that are turning a profit.

Posting a table containing data on his four selected programs (Figure 15), Dr. Nicholson noted that with the sole exception of IRAP, founded in 1962, they date from the previous decade. Still, all have enough experience that conclusions about their performance can be drawn. Pointing to the column that presents amounts of federal funding, he called attention to the fact that the Foundation for Innovation has invested the largest amount at $3.1 billion, and that the university research granting councils distribute just under $1.7 billion a year.

Canada Foundation for Innovation (CFI): Building Leading-Edge Infrastructure

CFI, the first of the four programs to be described, was set up to cofund leading-edge research infrastructure in universities and hospitals in response to a precipitous mid-1990s decline in the quality of research infrastructure in

Canadian universities—"another evil consequence," in Dr. Nicholson's words, "of the fiscal problem that the country was mired in." This foundation, which is government endowed, receives a grant worth a total of $3.1 billion in two or three installments; of that, it commits $2.5 billion to fund 4,000 projects through competition-based awards that are limited to 40 percent of project costs. In addition, there are special funds to encourage first-time researchers—a part of the program that has proved to be "extremely popular and high-leverage." There are also funds to help operate facilities, a "critically important" function because many advanced infrastructure components require highly technical operational assistance. Finally, funds are available to finance international collaboration. CFI's board, although it includes some government appointees, operates at arm's length from the government.

The objectives of the Canada Foundation for Innovation are fourfold:

- to transform research and technology development in Canada;
- to foster strategic research planning in universities, which Dr. Nicholson called an "interesting objective" and said had been "brilliantly achieved";
- to attract and retain world-class researchers, something that was taking place as well, as he would demonstrate with figures in a moment; and
- to promote collaboration and cross-disciplinarity, which he called "a huge success" of the program.

"The bottom line," he declared, "is that if you want to work at the leading edge, you need to have tools at the leading edge."

Fostering Strategic Research Planning

CFI's selection process begins with proposals that come almost exclusively from universities and research hospitals. These proposals are required to fit with an institution-wide strategic research plan at the applicant institution. Dr. Nicholson described the evaluation criteria as "fairly predictable"—research quality and need for infrastructure; contribution to innovation capacity building; and benefits to Canada. In a two-tier review, applications are grouped by subject matter and sent first to experts in their specific field, then to a multidisciplinary committee with the ability to make cross-area comparisons.

This selection process has forced institutions to develop long-term research plans and set priorities, in some cases for the first time. Its most significant consequence, in Dr. Nicholson's view, is that it has improved the quality of applications. Proposals worth $1.4 billion, a large percentage of them "absolutely top-drawer," were presented in the latest selection round even though the funding available covered only one-quarter of that sum. Every round, in fact, has raised the bar, with projects becoming larger and more complex. "While you would have expected that, once the low-hanging fruit was picked, the program might start running out of energy," he commented, but "exactly the opposite has happened." Collaboration among institutions has multiplied, research excellence has been stimulated at smaller universities as well as large, and most provinces have been led to establish similar programs at their own level. The latter are allowed to contribute to the 60 percent of complementary funding required by CFI, which has produced a "huge multiplier effect" within Canada's innovation system and has started to "put Canadian research opportunities on the world map."

A Virtuous Circle: Collaboration and Upgrading

CFI has "triggered a virtuous circle of collaboration and upgrading," Dr. Nicholson said, asserting that "strength begets strength." But, at the organizational level, he added there are three other lessons:

• CFI's arm's-length institutional structure and governance make its decisions credible to the point of being virtually unassailable, "not an easy thing to do in modern democratic systems."
• The endowed-foundation model has enabled CTI to plan beyond annual budget cycles. "They know how much money they have," he commented. "They always want more, but at least [they] know what [they're] playing with."
• The incentives are right, so that if the objectives of the foundation are met, further government funding is expected. Such funding is "not necessary, because this is an endowment," he said, "but the board and the people who run this program understand what they have to do."

Still at issue, however, is the level of investment that is appropriate for CFI. "At what point do you stop?" he asked. "Or do you keep following this bootstrapping process to higher and higher levels?"

Canada Research Chairs: Complementing Infrastructure with Talent

A human-resource complement to CFI is the CRC, whose objective is to develop a cadre of world-class researchers to exploit the infrastructure built up under CFI. CRCs' 2,000 chairs support all subject domains; 1,400 are filled to date. The potential number of chairs allocated to each university depends on the proportion of research grants it wins in other national competitions, although a bonus is reserved for smaller institutions.

The CRCs are divided into two tiers. The first, for world leaders in their disciplines, is an award of seven years' duration, renewable indefinitely, at $170,000 in support per year; winners in this category have access to numerous other kinds of support as well. The second, for "exceptional young faculty," provides support for 5 years at $85,000 per year that was renewable just once, for another 5 years. Under the selection process, universities are expected to nominate candidates for the CRCs in line with the same institution-wide strategic research plan to which CFI applications need to conform. Winners are selected by a three-person review panel or, in the absence of a consensus among panel members, by a standing interdisciplinary adjudication committee. The approval rate has been running at 85 to 90 percent of submissions, and acceptance by those approved has been around 95 percent.

CRC's outcomes have been extremely positive. At first, universities tended to nominate their best faculty in hopes of keeping them—although the program did not, as it turned out, inspire significant poaching among institutions. The focus has since shifted to recruitment, evidenced by the fact that, more recently, about 40 percent of chairs have been awarded to nominees from outside Canada. The relationship between the winning of research grants and the potential number of chairs has increased general interest in both; more specifically, the CRC program has significantly improved the research capacity of smaller universities. Together, CRC and CFI have "powerfully boosted Canada's research capacity at the front end," Dr. Nicholson stated. The only cloud over CRC's success has been the underrepresentation of women, who hold only around 15 percent of chairs reserved for world leaders and 20 percent of chairs overall. "To some extent that may be a factor of age demographics still," he speculated, adding, "It's something the program managers are very concerned about."

Assistance to Industry: IRAP

Moving from support for basic research to support for industry, Dr. Nicholson next described the IRAP, which he called "the classic one in Canada for small-business innovation." Its funding, about $135 million a year, underwrites the activities of 260 Industrial Technology Advisors operating from 90 sites across the country who are available to all small businesses engaged in technology. About one-third of IRAP's budget goes to consulting advice, which absorbs about 50 percent of these specialized advisors' time; about two-thirds, 30 percent of which is subject to repayment, goes to project support. The scope of project support is quite broad, encompassing everything from feasibility studies and pre-competitive R&D to international sourcing and youth hiring.

Of 12,000 clients served annually through the program, only 20 to 25 percent receive funding; the average level of such support is $30,000 per year, the maximum $425,000. Still, those 3,000 projects have encouraged formation of a network of 1,800 subcontractors, among them suppliers, consultants, universities, even the National Research Council. Dr. Nicholson called project approval "very fast"—2 weeks for projects under $12,000, up to one month for those up to $85,000. While pronouncing himself cautious about accepting existing evaluations of multipliers for IRAP spending, he quoted estimates as high as $12 in downstream investment for every dollar of spending on IRAP help, and up to 20 to 1 in sales or an equivalent. "The main point here," he said, "is that this is a real body-contact sport: These are people all over the country engaged face-to-face with small businesses."

One lessons from IRAP is that cost recovery for technical advisory services is unlikely to work. "Companies will not pay for what they don't know," said Dr. Nicholson, relaying wisdom from the program's director. Other noteworthy observations follow:

- The consulting industry values IRAP advisors as prescreeners and referral agents and, seeing its relationship with IRAP as symbiotic, has not complained about the program.
- The selection of sectors and topics of programs at this level needs to be client-driven.
- Proposals are better assessed as and when they are received rather than in request-for-proposal batches, as the former procedure fits with the firm's innovation cycle and also avoids peak-load processing problems.
- More focus is needed on mid-sized SMEs, those with 100-500 employees.
- Increasingly, SMEs need to be connected with national and global innovation networks.

TPC: Risk-Sharing and "Repayability"

The purpose of Technology Partnerships Canada, which has functioned as the Defense Industry Productivity Program until 1996, is to risk-share industrial research and precompetitive development across a wide spectrum. Designed to address a "persistent and frustrating" gap in Canadian firms' development of new technology, it covers from 25 to 30 percent of the costs involved in R&D, development of prototypes, and testing. In addition to significant cofinancing by industry, it features repayability, which depends on results. Targeting firms of all sizes and partnering with IRAP for SMEs, TPC parses its activities into three rather broad sectors: aerospace/defense, which, if considered separately, raise the number of focus sectors to four; environmental technologies; and "enabling" technologies, including biotech, materials, and ICT.

Only the program's budget, about $250 million per year on average, imposes a limit on the size of individual projects. From 1966 through 2004 about $2.3 billion has been committed, a little over one-third of that amount to SMEs—which, however, account for 90 percent of the 673 projects supported. Three-fifths of all funding goes to the aerospace and defense sector. Dr. Nicholson said that, according to "fairly careful calculations," private-sector recipients have matched TPC investments at an average ratio of 4 to 1. Project selection, which at 12 to 18 months takes much longer than in the case of IRAP, begins with the screening of applications, usually taking the form of "skeleton outlines," by in-house experts. Winners at that stage are invited to submit full proposals, which become the object of detailed due diligence. Negotiation then follow, where repayability and intellectual property provisions are among the major items covered.

Criticisms from Public and Participants

Dr. Nicholson then identified a number of "conundrums" raised by TPC that he judged "fairly characteristic" of such programs. Chief among these, perhaps, are those centering on repayability. This feature has been "oversold a bit," creating expectations among the public that have not been met. "Only about 3 percent of the funds out the door have actually come back so far," he stated. "But the truth is that we knew there was a long lead time for the repayment to come back; and, in fact, the program managers claim that repayment is on schedule." Compounding the issue, designing repayment terms that properly reflect the program's risk/reward-sharing component has been very difficult, since it was usually impossible to track what portion of a company's returns has accrued exclusively to a project supported by TPC.

The selection process has been another source of criticism. For one thing, project approval is protracted. For another, a perception exists in the political arena that TPC has been too focused on large companies—in particular, those in the aerospace sector—although the great majority of awards have in reality gone to SMEs. In addition, program objectives are so broad that it is difficult to maintain a consistent approach—or, put another way, there are so many grounds for approval that it is sometimes difficult to justify turning a project down. "That tends to invite a lot of objections from people who were disappointed," Dr. Nicholson said, "because someone can always find a precedent and say, 'But you approved that one, so what's wrong with me?'"

"What Does It Mean to Capture National Benefits?"

As "probably the most fundamental question" arising from the TPC experience, however, Dr. Nicholson cited the following: "In a world of global supply chains, what does it really mean to capture national benefits with programs like this?" Observing that many of the program's customers are multinationals—past TPC recipients include IBM, Pratt & Whitney, and RIM—he declared that it was "not quite clear that one can capture the benefits [for a national economy] to the extent that one once could."

Summing up, Dr. Nicholson expressed the opinion that Canada has rapidly built a strong basic research capacity that is paying off in terms of reputation and reversal of the country's brain drain. Its technology-development programs have evolved from subsidy-oriented to more sophisticated risk-sharing models, recalling a similar evolution in Finland. Finally, the lesson of Canada's experience, and one that he had seen reiterated throughout the morning's talks, is that any national innovation strategy today has to be globally linked.

DISCUSSION

Having commended the speakers for being on target in their presentations, and noting that he had taken over as moderator from Mr. Knox, Dr. Wessner called on Mark Myers to open the questioning.

The Special Demands of German Integration

Dr. Myers asked Dr. Kuhlmann to assess innovation programs in Germany since reunification from the time perspectives of the former East and West, as well as to look at economic growth in eastern Germany and to compare it with that seen in Poland, the Czech Republic, and Hungary. He expressed a specific interest in the equalization of salaries that took place right at the moment of unification in Germany, remarking that the other countries enjoyed, during at least part of their own restructuring, the "advantages of lower wages."

Dr. Kuhlmann replied that the question had been debated in Germany as well, but that historical circumstances had left no alternative. The East European countries, having become EU members, wcrc gradually availing themselves of the opportunities offered by "a European research system that is growing together." Salaries there had been relatively low 10 years ago but were slowly rising as those countries integrated with the European network. Thanks to this integration, he added, Eastern Europe can expect to develop advanced national research infrastructures in the longer term.

Germany's path had been different. Sudden integration caused a breakdown of the existing East German research system, which had not been compatible with that in the West. A marked brain drain resulted from salary discrepancies between the two regions: Especially early on, industrial researchers from eastern Germany found work with companies in western Germany and made their way there. Eastern Germany, meanwhile, was (and still is being) kept alive by huge public investments, without which there would be inadequate employment.

Innovation Policy—Also on the Demand Side?

Egils Milbergs of the Center for Accelerating Innovation, observing that most of the commentary from all three panelists had concerned enhancing innovation inputs—"more R&D, more scientists and engineers, more capital, etc."—suggested that a full discussion of innovation policy should include the demand side as well as the supply side. He urged consideration of macroeconomic policy's impact on innovation, specifically the role of interest rates; how tax policy affects the demand for innovation; standards; trade policy and how nations integrate with global markets; procurement policy; and competition policy. "It seems like there's an entire domain with huge influence on innovation that people who talk about national innovation systems don't really address in a meaningful way," he asserted. "They give recognition to these factors, but when you look at what the policies are, it's all about enhancing input, not maximizing output."

Canada: Political Obstacles on the Demand Side

Dr. Nicholson registered his agreement, saying that in his general policy role he probably spent more time on the demand side than on the supply side. But he allowed that, with the exception of tax cuts, the supply side is more visible politically because of the complexity of many demand-side issues, trade policy and competition policy being examples. Having spent most of his career in the private sector, he personally viewed competition as paramount: "Necessity is the mother of invention, there's no question about that," he said. "At the companies I've worked in, we were at our most inventive when the competition's breath was at our heels." This reality had been particularly salient as competition, once largely domestic, became global; finding a response to what was both a new source of demand and a challenge would occasion "a lot more head-scratching in all governments."

Against this backdrop, Canada has worked hard on its macroeconomic fundamentals, first getting control of monetary policy, then moving to fiscal policy. Amid massive tax cuts, its corporate tax rate is now down to below that of the United States. More recently, under a "smart regulations initiative" fueled by recommendations from a blue-ribbon panel for improvements in such areas as drug approval, the country has started to eliminate "the little differences that don't make a difference" but nevertheless interfered with trade across the U.S.-Canada border. Another area ripe for streamlining—admittedly, "a very danger-ous word"—is environmental regulation. But while ample "low-hanging fruit" in the regulatory area offer ways in which "government policy [could] be helpful on the demand side of the equation," he said, "none is easy politically."

Germany: Trying To Go Beyond Classic Recipes

Dr. Kuhlmann also pronounced himself in agreement with the statement that far more elements of public policy have an impact on innovation than those included within the "narrow notion of innovation policy." As understanding of this area grew, European governments were more frequently becoming the object of criticism that their innovation programs did not take account of broader economic issues. Those formulating innovation policy, meanwhile, are finding themselves obliged to design their initiatives in the context of other policies whose impact on "what actually happens in companies" might be greater by far. Discussion of these questions is increasing in numerous European countries, Germany among them, although the debate is limited to experts rather than taking place among the general public.

Former Chancellor Schroeder's Partnership for Innovation was an attempt at a systemic or holistic approach to influencing the basic conditions of innovation that went far beyond classical policy instruments. But, speaking from his experience with governments at different levels in Germany, Dr. Kuhlmann saw a problem in the prevalence of competition among agencies. Unlike some competition, this variety is not productive; rather, being rooted in claims by each agency that its work was more important than that of the others, it results in gridlock, "quite a mess from the perspective of innovating companies." Some degree of coordination, including some effort at mutual information and collaborative design, is therefore necessary.

In Finland, perhaps, designers of demand-oriented innovation policies ask "What is the problem, and what can we do about it?" rather than occupying themselves with the borders among ministries. Only at the end of the design process might they ask, "Who is in charge of what, and how would we have to redesign structures?" This sequence does not, however, represent the norm in public policy making in Germany.

Cross-Border Comparisons of Attitudes, Programs

Dr. Wessner then posed three questions that, although quick in the asking, he judged likely to demand lengthy responses. He requested that Dr. Nicholson, whom he complemented on a superb presentation, discuss how the Canadian government had dealt with what sounded like charges that some of its innovation programs were vehicles for corporate welfare that funneled the majority of available funding to large companies. Referring to the STEP Board's assessment of the U.S. Small Business Innovation Research (SBIR) program, which was in progress, he described himself as having been heartened to hear that someone—in this case, Dr. Nicholson—shared his acquaintance with the difficulty of ascertaining the consequences of a single award for the future of a company.

He then asked Dr. Kotilainen whether he regarded the budget of the United States' Advanced Technology Program, which had been running at around $140 million per year, as adequate to the U.S. economy in light of the fact that Finland, a nation with 5.1 million inhabitants to the United States' 300 million, was spending between $530 million and $560 million annually on a very similar program.

Finally, Dr. Wessner said he had observed that while Europeans enthused over the creation of a European research space and innovation system, they often had barely concluded when they began speaking about their own national programs. He asked Dr. Kuhlman to talk about those two elements in relation to one another, then very briefly to discuss the impact of the EU programs, a subject Americans were very interested in, and to assess whether innovation efforts in Europe are primarily driven at the EU or national level.

Parrying the "Corporate Welfare" Charge

Dr. Nicholson said that, as charges of corporate welfare are still to be heard, the problem that Dr. Wessner alluded to is yet to be solved. Because subsidies to business have been cut "pretty dramatically," however, he felt that "just in terms of the gross dollars" the critics could no longer make as strong a case. Canada's federal budget surpluses had also ameliorated the situation, by rendering the trade-offs less obvious. A more substantive change, the move to risk-sharing with repayability as featured in the Technology Partnerships Program, had placed a new obstacle before opponents of the programs, although they nonetheless took the line: "You haven't gotten all your money back yet, so obviously this must still be corporate welfare." Also helpful is that, increasingly, potential criticism is arising in the context of a local facility where a large number of jobs are at stake. Then, "frankly, the political shoe is sometimes on the other foot," Dr. Nicholson said, for "if there isn't a little bit of big-company support, there's a larger price to pay."

Administering Supply-, Demand-Side Policies Separately

Dr. Kotilainen began his response by taking up the issue of demand-side measures in support of innovation that Drs. Nicholson and Kuhlmann had previously addressed. At Tekes, he said, innovation policy is viewed as far broader than most other policies implemented by the government, including science policy, technology policy, and industrial policy. And because it incorporates elements of all the others, innovation policy is difficult and complex to run. "Therefore, we think we should concentrate on that," he said, "leaving the other policies to the private sector or to other parts of the government to take care of." Even though Tekes finances research, the main focus of its awards is always industrial competitiveness and networking, a strategy that leaves the rest to the companies themselves.

Addressing the matter of the Advanced Technology Program (ATP), Dr. Kotilainen emphasized its philosophical similarity to Finland's national innovation programs. Those programs have been extremely beneficial for the country because they combine the skills of its universities with those of its companies. Projects are planned jointly by the universities and industry from the very beginning, so that both know exactly what to expect from the research, and companies also got acquainted with university researchers, a very good basis for subsequent recruiting. This, he remarked, is an essential function no matter what the size of a country is. On the specific subject of ATP's annual budget, Dr. Kotilainen recommended that, rather than being cut, it should be favored with the addition to the end of a zero or two.

Balancing Europeanization, National Programs

Dr. Wessner, announcing that the luncheon speaker had arrived and that the session would thus have to be concluded quickly, asked Dr. Kuhlmann to be brief in his remarks on the relationship between Europeanization and those national programs still in place within the EU.

Dr. Kuhlmann said he had speculated in a paper published just two weeks before on how "this very contradictory relationship," so fraught with tension at that moment, might develop in the future.[10] There were some government officials who are opening up to collaboration across borders; there is, in fact, a program called ERA-net that funds only those intergovernmental collaborations aimed at developing joint national programs on a European platform. At the same time, however, other officials still fail to see the value of such efforts, whose very existence, at least partially, call into question their role as national policy makers.

---

[10]J. Edler and S. Kuhlmann, "Towards One System? The European Research Area Initiative, the Integration of Research Systems and the Changing Leeway of National Policies," *Technik-folgenabschätzung: Theorie und Praxis*, 1(4):59-68. Accessed at *<http://www.itas.fzk.de/tatup/051/ edku05a.pdf>*.

Luncheon Address in the Great Hall

Moderator:
William J. Spencer
SEMATECH, retired

Dr. Spencer, expressing his honor at introducing Dr. John Marburger, pointed out that the latter's current tenure as science adviser to the President and director of the Office of Science and Technology Policy (OSTP) coincides with "an exciting time for science in the United States." Moving selectively through the details of Dr. Marburger's extremely distinguished career, a printed summary of which was available to the audience, Dr. Spencer represented him as an exemplar of a generation that grew up believing it was physics that underlay all science. After earning a master's degree at Princeton and a Ph.D. at Stanford—"two of the better physics departments in the country," Dr. Spencer pointed out—Dr. Marburger distinguished himself as a scientist in the field of nonlinear optics.

But what gained Dr. Marburger the most recognition, even before his elevation to his current position, was his leadership in the field of science. He had served as chairman of the Department of Physics at the University of Southern California, then as president of the State University of New York at Stony Brook. When, in 1998, Dr. Marburger moved into the post of director at Brookhaven National Laboratory (BNL), Dr. Spencer was a member of BNL's board of directors; he therefore knew first-hand "what a mess" the laboratory was in at the moment Dr. Marburger took it over.

As Dr. Marburger had consented to take questions following his address, Dr. Spencer decided, in the interest of time, to turn the podium over to him without further ado.

LUNCHEON ADDRESS

John H. Marburger
White House Office of Science and Technology Policy

Registering his appreciation at having been invited to address the symposium, Dr. Marburger said that its agenda struck him as strongly similar in theme to that of a conference he had addressed a month before: "Innovation as a Competitive Advantage: Role of the Research Park," sponsored by the Association of University Research Parks (AURP). He decided, therefore, to present substantially the same remarks as on that previous occasion.

Research parks became a global phenomenon and, in the future, are likely to be significant focal points for all countries with knowledge-based economies. In 2002, the AURP identified and sought data from about 200 research parks associated with universities in the United States. Those that responded, about half of the 200, had a total of more than 2,900 tenants employing over 235,000 individuals; the dominant or leading technology was, in most cases, biomedical or medical technology. Not-for-profits made up 83 percent of those responding, while 62 percent had a business-incubator component and about one-third had a yearly operating budget exceeding $1 million. Seventy percent had been established using public funds, mostly during the 1980s and 1990s. Dr. Marburger acknowledged that these data were a few years old, but opined that they were not of a sort susceptible to rapid change. The statistical portrait emerging from them matched the characteristics of the technology park and incubator programs that he started at Stony Brook in the 1980s and early 1990s, when he was the institution's president. His remarks, therefore, would be based on direct experience with the development of such parks.

Throughout his career, Dr. Marburger said, he had asked himself: What is the best strategic path to successful technology-based economic development? Although he did not give so much thought to this question while he was busy solving problems of quantum electrodynamics as a graduate student in the early 1960s, it was nonetheless in the air. Everyone at Stanford in those days was aware of the Stanford Research Institute, soon to become SRI and SRI International; of the rise of nearby Silicon Valley, which was just coming together; and of how the university had planned to foster high-tech industry even before World War II, an idea finally being carried through. "For those of us who worked in the Stanford environment, even if only briefly," he recalled, "the power of linkages between a major research university and regional business was immediately obvious." Another university-based coalition had emerged in North Carolina just a few years earlier to take advantage of the intellectual assets of Duke University, North Carolina State, and the University of North Carolina at Chapel Hill. It had an institute, RTI, which was similar to SRI. Also like Silicon Valley, the Research Triangle initiative had originated in academia and quickly engaged regional business leaders.

In these two examples of partnerships between universities and business, the federal government had been an important silent partner. The expansion of engineering and science departments during World War II, along with rapid increase in federal funding for university research in the decades thereafter, had created centers of excellence in basic and applied research throughout the country. State and local governments were making important investments in higher education—particularly conspicuous were those of California and New York, which were building new campuses—and 2-year community colleges were springing up everywhere. The recruitment of an increasing percentage of high-school graduates into new post-secondary programs, above all those of the 2-year colleges, was taking place. Governments helped build an infrastructure in which the new partnerships could succeed. "My colleagues on the President's Council of Advisors on Science and Technology [PCAST] during this administration refer to this whole phenomenon as 'the ecology of innovation,'" Dr. Marburger said, explaining that the term suggested "the notion of a 'system' in the innovation process." This has been a subject of much study; in fact, he reckoned that many attending the symposium have looked at economic development occurring around the nucleus of a research university or federal laboratory, where this development has been fostered by the investment of public funds at the federal and state levels.

Five categories embracing institutional participants in innovation systems were identified in a study prepared for PCAST, which Dr. Marburger cochaired with the prominent Silicon Valley venture capitalist Floyd Kvamme, by the Science and Technology Policy Institute (STPI), an organization attached to OSTP and operated by the Institute for Defense Analyses:

1. governments, which play a key role in setting broad policy directions and a primary role in funding basic scientific research;

2. private enterprises and their research institutes, which contribute to development and other activities that are closer to the market than governments are;

3. universities and related institutions that provide key knowledge and skills;

4. bridging institutions acting as intermediaries under such names as "technology center," "technology brokers," or "business innovation centers," which play an important if understated role in closing the gaps among the other actors and had been important to the success of all types of research centers; and

5. other organizations, public and private, such as venture capital firms, federal laboratories, and training organizations.

The STPI study also identified four contextual factors sufficiently important to the operation of innovation systems to be able to make or break them: market conditions, physical infrastructure, education and training, and regulatory conditions. The innovation-systems approach looked at by PCAST the previous year identified barriers to specific policy objectives and assigned functions to mitigate them for each of the five categories of actors. The process, as Dr. Marburger summarized it, was one of sorting through the challenges, deciding which actors should address them, and systematically improving the environment so as to promote the success of research centers.

As an example, he offered the policy objective of building an innovation culture where none had existed; this might be addressed by removing regulatory or legal barriers, which is a government function, and providing incentives for venture-capital funding, which government and private enterprise can do together. Enhancing technology diffusion among the actors, then promoting extension and technical-assistance programs, is a role of government and bridging organizations, while government also can promote lifelong learning with the help of universities and private enterprises. Taking a systems approach that identifies the actors by functional categories in this way and working systematically through issues in the contextual factors that might be inhibiting the development of these centers turns out, Dr. Marburger said, to be an effective strategy. "Government can play a role in most strategies," he added, "but it is not the only actor, and generally government action is most effective when it responds to needs identified by actors that are closer to the market."

Returning to the subject of his early exposure to regional development, Dr. Marburger recounted that, upon finishing his graduate study at Stanford, he started his research career at the University of Southern California. There, too, he saw first-hand how the engineering and business schools, strongly committed to regional industry, were leveraging federal funding to build programs that supported regional development. While there, he himself started a center for laser study with federal and industrial cosponsors that 30 years later was still thriving.

He took these lessons with him when he went in 1980 to Stony Brook, where his first task as president was to open a 540-bed tertiary-care university hospital. In the process of "learn[ing] a lot more than [he had] ever wanted to know about the hospital business," he saw an opportunity to leverage the resources that the state of New York was willing to invest—in both the hospital and expansion of the university's medical school—to build up bioscience research, health care, and regional industry all at one time.

Long Island, where Stony Brook is located, had multiple assets for developing biotechnology at that time. The director of Cold Spring Harbor Laboratory, the Nobel Laureate James Watson, was concerned about his scientific staff leaving the region to start biotech businesses in California and elsewhere. Nearby Brookhaven National Laboratory also had important resources for biomedical research although it was almost totally disconnected from regional industry. And the university was about to expand its medical faculty dramatically, not only in the clinical departments but in the basic medical sciences as well.

Running through the highlights of what he said was a long story, Dr. Marburger related that Stony Brook established a degree program in genetics with Cold

Spring Harbor lab and placed one of its new hospital's three linear accelerators for cancer therapy at Brookhaven. The university's dean of medicine became a member of the same BNL board of directors on which Dr. Spencer was serving. The three institutions began joint recruitment with the aim of building complementary competencies. And Stony Brook emerged from a New York state competition set up to fund centers for advanced technologies with an award for one specializing in medical biotechnology.

At the same time, the university began building acceptance among its faculty for participating with industry in cooperative research programs. "There was a lot of controversy at the time about whether engaging in industrial-related research would somehow undermine the quality of academic programs or the purity of the research," Dr. Marburger recalled, adding: "You don't need the entire faculty to have an attitude that's favorable to development, but you do need a critical mass."

These and related activities strengthened Stony Brook's applications to the state economic development agency for low-cost financing and grants to build a biotechnology incubator facility next to the hospital, which became a huge success. High occupancy rates right from the beginning made it possible to maintain an aggressive business plan for the incubator and led to several rounds of expansion that eventually included the construction of a generic pilot manufacturing facility for biotechnology tenants.

The university also developed a small technology-transfer office that worked quite well, probably because its director, a former small businessman, had had direct experience with the commercialization of technology. He spent his days walking around to the labs, knocking on their doors, and asking people what they were doing. "If it sounded interesting to him," recalled Dr. Marburger, "he took out the papers for patent disclosure and said, 'Here, I'll help you.' And if it didn't sound interesting, he said, 'Well, it's yours, do with it what you want.' " His activities led to an exceptionally high rate of licensing per patent disclosure.

The director of the technology-transfer office was associated with a person outside who spent his time shopping around a list of university intellectual property to customers throughout the world. The two helped Stony Brook develop a very strong and successful business in the wake of the 1980 passage of the Bayh-Dole Act that had remained a source of revenue for the university and of satisfaction to many on the faculty, who had in the meantime become much more comfortable in their engagement with industry. But in recognition of their success, both the director and his colleague were promoted into other positions, with the result that the university's technology-transfer operation "got bigger, and much less efficient and effective," Dr. Marburger said. "It is still functioning today, but it never quite functioned as well as when we had only two people and a secretary."

All these activities established links between the university, regional business, and government economic-development agencies. Coordination among the major research institutions allowed them to build complementary strengths and present a broader interface to the business community. The creation of critical masses of talent in related fields served to raise mutual awareness of opportunities and to reinforce the directions that the institutions had already established in their long-range plans.

Looking back on these experiences, Dr. Marburger isolated five principles that he viewed as key to success:

1. **Build competencies with attention to regional strengths.** This is important for a large country like the United States, whose markets display very strong regional differences but each of whose regions has its strengths and its possibilities. Institutions cooperating in regional development must hire people whose interests enhance and complement what is already found in the environment, which "doesn't happen unless somebody pays attention to it." For the idea is to build *regional* strength, not just *institutional* strength. When several research institutions are located in the same region, they benefit by cooperating in recruitment and group development. Stony Brook, Cold Spring Harbor lab, and Brookhaven National Lab shared information on an informal basis about areas of concentration and often collaborated on recruitment.

2. **Identify a research strategy.** Stony Brook's conscious decision to make biomedical research a priority meant allocating university resources to proposals and projects that worked together to build a foundation for future successes—even if, "in terms of some sort of absolute measure of quality," these were at times not the best proposals to come forward. While there were exceptions to this practice, a bias was maintained in favor of those fields that could be expected to help further the overall strategy. "That requires leadership," Dr. Marburger declared. "It does not happen in a university environment unless someone is willing to push on it." Faculty development and capital improvements were coordinated to enhance biotechnology capabilities. While other areas needed and deserved attention, the immediate opportunities for funding lay in the biosciences, which therefore received the focus.

3. **Build a regional environment.** In the early 1980s, Long Island business organizations were not aware of the rapidly growing opportunities in the biotechnology industry. They did not appreciate the significance of an emerging major tertiary health-care facility or the value of federal funding as a source of technology. The Long Island economy was then dominated by large aerospace contractors—principally, Grumman Corporation—that were to fall by the wayside as the cold war came to an end and industry shifted completely. "So it was important for me and my counterparts at the two laboratories to get together, pound the pavement, and talk to people—to take the biotechnology message to business groups, chambers of commerce, and state and local government agencies," Dr. Marburger recalls. "The whole region had to cooperate in making this work, and somebody always has to take the first step to get others together." Because Long Island's

business community was aware of the dangers of relying on a single industry, these efforts by the leading centers of research to work together with business were warmly received.

4. **Form regional partnerships.** Institutional rivalries are counterproductive; cooperation and collaboration are essential for regional-scale development; and regional-scale development is important for a stable pattern of growth. That companies start up, grow, then frequently either die or move elsewhere is not necessarily the end of the world, but it does necessitate continual start-ups. Some of the new companies may survive and add permanently to the economy, some may have to be replaced with others that are sufficiently similar to stabilize the workforce. It is because regional partnerships enhance mobility and multiply opportunities for workers and for businesses that a critical mass of mutually compatible businesses is needed to stabilize the inevitable effect of start-ups' moving away. "In Silicon Valley in its heyday, and it is presumably still somewhat like this, you had the phenomenon of frequent moves of technical personnel from one company to another," Dr. Marburger observed. "There was a great deal of mobility—companies came and went, started and failed—and in general the makeup of the workforce was similar, which stabilized employment in the area despite the dynamics in the companies."

5. **Fund the machinery, which consists of facilities, people, and organizations.** None of this happens without people who know that their job is to make it happen; neither regional development nor technology transfer can be made to work with volunteers. "I travel around the country looking for regions that are succeeding," Dr. Marburger said, "and many are attempting to do it on a voluntary basis, but only those where there is some sort of executive center with a paid workforce [are having success]." In other words, whether at a state, county, or local-government economic-development office, or at an organization that is either freestanding or associated with a university or a business group, someone has to know that technology transfer is his or her job. Technology-related economic development usually entails investing state- and local-government funds in facilities to reduce costs for start-up tenants. Also needed are people to bring entrepreneurs together with financial and technical support. More than brokers, these individuals play the role of teacher and counselor for entrepreneurs who know the technology but are not familiar with business practices, and for investors who are not familiar with engineering and scientific mindset.

Dr. Marburger acknowledged that these lessons may be learned in other contexts and may not apply to every situation. He and his Long Island counterparts tried to learn from other regions as theirs grew, and they discovered national organizations that help to foster best practices. The movement that supports university-based research parks, and of research parks that are based around a nucleating asset other than a university, was growing, thriving, and becoming an important part of the U.S. innovation ecology.

Dr. Marburger reiterated in closing that his observations were based on his own experience in addition to his knowledge of other research parks in the United States and around the world. He added that over the previous 5 years, he has visited over a dozen research parks in South Korea, Japan, Russia, and Europe. While "impressed with the similarities among the regions," he noted that research parks abroad have a "much stronger government component" than is found in the United States. However, he said that he perceived the sheer diversity of such enterprises found in the United States, implying both experimentation and potential adaptability of research parks to new realities and opportunities, to be a source of strength.

DISCUSSION

Dr. Spencer, thanking Dr. Marburger for his comments and concurring that some leadership is required to set a research strategy, asked him to provide guidelines for doing so, particularly in an environment where researchers are unused to working in accordance with a strategy, such as that of a university or national laboratory.

Dr. Marburger answered that among the first things to be done is to gather data. He had seen groups, whether they were made up of faculty or of university presidents, get together and have a great idea but make the mistake of not having "a look around" to explore possibilities. They may then see their efforts to sell their project meet with disappointment because it does not seem to fit with anybody else's plans, whereupon they may either attempt to implement it alone or get a federal grant that in the end builds up a lab or center within the university that grows no roots in the community. Some states, as they undertake their own economic-development plans, have contracted with research organizations such as Battelle to make surveys of the capabilities found in their regions, and have then based their economic-development plans on the resulting data. The state of New York has very consciously selected, based on actual studies of capacity resident in the state, a set of objectives for fields of research and has promulgated it through its economic-development operation.

Consensus building—which starts with procuring regional support and extends to creating an environment of acceptance for moving in a certain area—is never easy. But starting with an idea that doesn't fit and then trying to force it is also very difficult. "You have to take advantage of opportunities," Dr. Marburger commented, "and this doesn't align very well with the values that you often find in research universities." When looking for a faculty member, a research university very often simply looks for the best person available even if his or her field doesn't quite fit its requirements; or, conversely, it might leave a position unfilled for a long time until it can find the person it really wants. "You've got to be a little bit more flexible than that if you're building a capability that will fit with a lot of other partners," he stated.

Next, Dr. Lonnie Edelheit, former senior vice president of Corporate R&D at General Electric, asked Dr. Marburger what he sees as the role of government in such enterprises. Since the kind of undertaking Dr. Marburger proposes demands integration, as well as management and leadership, he asked: "Is there enough of that in the government, and where should it be?"

In his response, Dr. Marburger noted that agencies like the National Science Foundation (NSF) increasingly fund theme-based programs—although they continue to be peer reviewed and merit based. For example, NSF funds centers that conduct competitions based around particular themes, such as materials research. In addition to this thematic focus, NSF also often requires these centers to cooperate with regional industry and state offices. Reflecting this stipulation, proposals from around the country often include testimonials from regional business groups and state and local government officials.

A tension exists within funding agencies between two desires: to have money in big, undifferentiated pots to respond to unsolicited proposals and, increasingly, to hold out a share for fairly well-defined programming in a field such as nanotechnology or information technology. At the agencies that fund extramural programs, Dr. Marburger said, a trend over the past decades of reserving more and more of the money for theme areas is discernible. "That's quite reasonable," he commented, "that's the sort of thing you want." There are currently some 27 nanotechnology centers being sponsored by four or five federal agencies, and OSTP tries to coordinate this "interesting phenomenon" by keeping track of what the centers are doing and by making sure that the agencies talk to each other about how the capabilities of the different centers fit with one another. Having peer-reviewed, merit-based grant awards is therefore consistent and compatible with the type of development that Dr. Marburger deemed effective.

Referring to the five points mentioned by Dr. Marburger, the next question concerned the environmental and health aspects of regional development.

Dr. Marburger explained that all such activity takes place in a societal context and that responsibility for governing its different parts lies with different authorities. To illustrate, he noted that environmental issues are usually subject to a licensing process, building permits, zoning, and so on, all of which are part of a complex of issues that have to be managed together. Typically, it is possible to set up a system of licensing and regulatory control over new business, especially in the high-tech sector. If hazardous chemicals or operations are involved, the system may possess "very antigrowth, very inhibitory" features, he said, adding: "That needs to be worked out with local authorities so that you can have responsible growth."

One of the more interesting phenomena that Dr. Marburger had seen on Long Island, where a very high degree of environmental consciousness existed, was a coalition of environmental-advocacy groups that had been formed to work with developers and the business community. "They realized that if you tried to hold back all development, then you'd get a fragmented, irresponsible, not very efficient, and environmentally destructive pattern of development," he recalled. "So the two sides sat down and worked out a pattern of development priorities," one result of which had been the Long Island Pine Barrens Initiative.

Such achievements are, therefore, within reach, but leadership and "a very careful weighing of the pros and cons" are indispensable. Some types of operations are simply not compatible with certain regions, Dr. Marburger warned: "You can't just plop down a manufacturing facility that uses large amounts of chemicals in an environmentally sensitive region without everyone understanding what's happening." So a role for government does indeed exist, but because the execution of social requirements is usually very distributed, it could not be carried out from the top-down. Accounting for part of the complexity is the fact that enterprises around the country are growing up in the context of environmental and health regulations administered at federal, state, and local levels.

A[n unidentified] questioner observed that size and resources afford the United States a very big advantage in a global economic environment that is growing much more competitive, but that other countries have achieved greater government/private-sector coordination. He asked whether Dr. Marburger thought that, in the new environment, it would make sense for the United States to have more of the kind of coordination seen elsewhere and how he thought it might be structured.

Dr. Marburger acknowledged "the focus that other countries are able to give to economic development," along with the achievements that focusing their resources has produced, as a "scary phenomenon." He asserted, however, that opting for such a strategy presumes an ability to foresee what would add the most value in the future. Noting that the huge diversity that the United States has and will retain is one of its advantages, he stated that he hoped that, in the end, the U.S. approach of relying on diversity and market-based planning would provide the country with a number of different models in a number of different areas. "Our size and experience in this, and the fact that we don't constrain all [our] institutions to the same model, makes us an ongoing laboratory," he said.

The fact that the country's research parks are growing up in what Dr. Marburger called "different governmental ecologies"—under different regulatory environments, in different cultures—impressed him, therefore, as a source of strength. He praised the efforts of such organizations as the AURP to communicate with each other and to engage in sharing best practices and similar activities. Conferences like the one he had addressed the previous month attract not only those who manage the research parks but also representatives of state and federal governments, who are watching developments. "We're getting educated," he stated, "and we have confidence in our rational ability to make things work."

It was because no one has an economic model that forecasts what would work in the future, Dr. Marburger stressed, that the simple fact of the United States' huge capacity place it in what would ultimately be a safer position. The one indicator that is always hopeful for the United States is the scale of its economy and the very large number of research centers it has. "But there's the rest of the world out there, and it is developing, and at some point we will just be a part of this developing world," he said.

Responding to a question from Gregory Schuckman of the University of Central Florida about forecasting future workforce needs, Dr. Marburger noted that, while the issue was both extremely important and interesting, there is no model as yet that is capable of predicting success in economic development or that can predict future technical workforce needs. The jobs of engineers, technicians, and scientists are changing just as rapidly as jobs in other sectors: Entire categories of technical work have disappeared during the decades of the information-technology revolution. In the same way that office productivity has increased—as measured, perhaps, in the number of people required to generate a certain number of pages—the science and engineering community have become more productive. But as "we do not have a very good handle" on the rapid changes in patterns of production and the relation of the workforce to production, he was skeptical about predictions regarding the workforce.

Trends could be observed. There was no question that every developed country in the world is concerned about its engineering workforce in light of developments in China, but the rate at which China appears to be outproducing the rest of the world is probably not sustainable. "We're not working in linear systems here," Dr. Marburger declared. "Those rates are going to turn over eventually, and I don't know what the forces are that will turn them over." How all those being trained in Asia would be absorbed into the technical economy was unclear.

What is known, Dr. Marburger said, is that understanding of the natural world in fairly sophisticated terms, and particularly technology, will be a part of our way of life in the future. It will be important for people to understand how the world works in some quantitative detail in order to have any job in any part of the economy. He pointed to this recognition as "one of the deep philosophical perspectives" behind the concept of No Child Left Behind: "We've got to make sure that our young people are adequately prepared for these futures." Whatever future jobs look like—whether in engineering, in a new form of business training, or in other fields—better quantitative skills would be imperative.

There was, however, some good news. The proportion of those graduating high school having taken a course in physics has been going up steadily in the past two decades to surpass one-third, compared with about one-quarter 10 or 12 years earlier. The same applies to precalculus: About one-fourth of current U.S. high school graduates took a precalculus course and the proportion has been rising steadily for the past 10 years. "Something is happening out there," said Dr. Marburger. "Somebody is getting the picture."

Still, many were flunking their first-year college courses in math and science and were becoming discouraged. "They came wanting to be scientists and engineers, and we turn them away," remarked Dr. Marburger, recalling that he himself had taught freshman physics and knew how tough introductory courses could be.

In the interest of producing more scientists, teaching methods need to be altered to take advantage both of the knowledge with which young people are in fact equipped with and of their desire to understand how the world works.

Dr. Marburger declared himself "pretty optimistic" overall. While conceding that one could point to quite a number of "scary indicators," he said that it was part of his job to worry about precisely what those indicators meant. As the President's Science Adviser, he could be an advocate or a counselor. While he acts as the former at times, there are times when he felt obliged "to sit back and say, 'What does this all mean?'" He said that he is working hard on this problem, which is complicated by the existence of conflicting analyses. Earlier the same morning, he had conferred with National Research Council staff engaged in a study of these questions, and they shared his concerns. All felt the need for a much improved framework for gathering, using, and analyzing statistics relating to the workforce and innovation. "So we'll keep trying," he concluded, "and I think trying is one of the most important things that you can do."

Dr. Spencer, ending the session, allowed that the news conveyed by Dr. Marburger about the number of high school students taking physics was the best thing he had heard up to that point in the day's meeting.

New Models in Japan, Taiwan, and China

Moderator:
Alice H. Amsden
Massachusetts Institute of Technology

Dr. Amsden opened the session by stating that, given the history of Japan, Panel III's examination of China, Taiwan, and Japan would offer a look at how innovation or technology systems differ from developed to developing countries. She said that in listening to the previous discussion of Finland and Canada, as well as to Dr. Marburger's luncheon address, she had been struck by how little was said about industry. In contrast, she noted, a consideration of China's technology policy always revolves around an industry; there, the theme might be "technology policy and how it develops the telecommunications industry," or, in the case of its neighbor, "the technology policy in Taiwan and how it developed the computer industry." These countries would, of course, be much more focused on industry because they do not yet have industry fully in place and so were catching up. The acquisition of technology is therefore very instrumental to their innovation policies, with such underpinning concerns as: How can we get the technology? Should we make it ourselves or buy it? How can we get the technology in order to make an industry competitive internationally?

Developing countries do seem to have stopped there. However, offering Taiwan and Korea as examples, Dr. Amsden noted that some are moving into a more exploratory area, where the relevant questions become:

• How can we create new technology that spawns new industries, or that allows us to introduce mature high-tech industries like thin-film transistor liquid-crystal displays (TFT-LCD)?

• How do we get the technology that allows us to produce these products and go around existing patents, or to compete in world markets, when this technology is changing so fast?

Suggesting that one of the differences between developing and developed countries (when it came to innovation or technology systems) might be the degree of emphasis on industry, she speculated that the panel's speakers might shed some light on the issue.

Dr. Amsden then introduced Dr. Hsin-Sen Chu of Taiwan's Industrial Technology Research Institute (ITRI). She characterized the organization as one of that country's very best research institutes and noted that Dr. Chu is well known for his clarity of vision and his interesting take on technology and development.

THE TAIWANESE MODEL: COOPERATION AND GROWTH

Hsin-Sen Chu
Industrial Technology Research Institute (ITRI), Taiwan

Dr. Chu said that it was his honor to describe Taiwan's model of and experience with technology development and innovation. Like Finland, he said, Taiwan is a country that is small in population and area whose resources are sufficiently limited such that they must be used with a maximum of efficiency and effectiveness. His presentation comprised five parts:

• a brief introduction to the transformation of Taiwan's economy over the previous 50 years;

• a discussion of the major elements in Taiwan's industrial evolution;

• a brief introduction to ITRI, covering the role it had played in that evolution and the nature of its relationship with Taiwan's industries;

• a description of the opportunities before Taiwan; and

• a very brief conclusion.

To begin, Dr. Chu projected a graph illustrating the change in the composition of the Taiwanese economy between 1951 and 2004 (Figure 16). The country's per capita GDP had grown from $145 at the start of that period to $13,529 by its end, a nominal increase approaching two orders of magnitude.[11] The service-industry

---

[11]Adjusted for inflation, the growth in GDP per capita in Taiwan over the period from 1950 to 2000 is likely to smaller, though still impressive 20-fold increase. See Angus Maddison and Donald Johnston, *The World Economy: A Millennial Perspective,* Paris, France: Organisation for Economic Co-operation and Development, 2001.

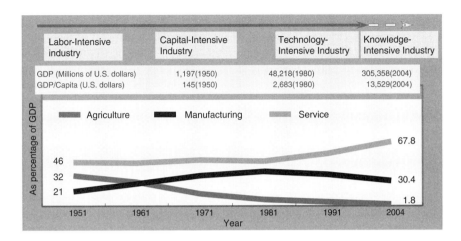

FIGURE 16 Transformation of Taiwan's economy.

sector accounted for about 46 percent of the economy from the 1950s through the 1970s, when a steady rise began that brought its share to 67.8 percent by 2004. Agriculture's contribution had moved in the opposite direction, dwindling from about 32 percent in 1951 to less than 2 percent currently. Manufacturing, meanwhile, displayed two contrasting phases: Its slice of the economy climbed from 21 percent in 1951 to near 40 percent by the end of the 1970s, only to fall back to 30.4 percent by 2004 as it was overshadowed by the continuously growing service sector.

Transformation of Taiwan's Economy

If technology-based manufacturing was the engine driving Taiwan's economic growth over the previous half-century, the country expects services to act as a "twin engine" in the future. Similarly, while labor-intensive industry characterized industrial growth in the late 1940s through the 1950s and 1960s, followed by capital-intensive industry the 1970s and 1980s and technology-intensive industry the 1980s and early 1990s, the knowledge-intensive industry that came onto the scene in the mid-1990s is expected to expand further in the future. Describing the economic evolution in Taiwan, Dr. Chu said that the food and textile industries emerged in the 1950s and 1960s, followed by bicycles, motorcycles and selected basic industries in the 1970s and 1980s. It was in the late 1970s that Taiwan's semiconductor and other information-technology industries started to take root. Optoelectronics made its appearance in Taiwan in the early 1990s.

Currently, Taiwan ranks among the top three in the world for products over a wide range of sectors. In the computer and peripherals category, it is the world leader in notebook computers, small and medium-sized TFT-LCD modules, as Dr. Amsden had mentioned, but also in both CD/DVD drives and disks. In network products, Taiwan is a leader in wireless LANs, hubs, and SOHO routers; in the integrated-circuit (IC) sector, it leads in the foundry, mask, IC design, and DRAM areas. And it retains some world-leading products in the consumer sector, among them bicycles, hand tools, and textiles.

The evolution of traditional sectors and the development of new industries can be illustrated by examples below.

Example One: Textiles

Charting a half-century of progress in the textile industry, Dr. Chu reported that, in the beginning, Taiwan simply imported such raw materials as cotton, wool, and silk for processing. The next step was to move from natural to artificial fibers and into the apparel industry while also developing the manufacture of textile-fabrication machinery. Over the previous 10 years, the government had supported R&D efforts that he characterized as "tremendous" in the areas of functional fabrics and industrial textile fabrication technology. Encompassing nanotechnology, these efforts have yielded many new products in recent years, some of them in the field of fashion design. In 2004 Taiwan was the world's sixth-ranked textile exporter with export revenues of $11.9 billion.

Example Two: Electronics

Taiwan is the world's fourth-largest IC manufacturer, having posted 2004 revenues of $33.3 billion. Its chip industry, begun when RCA transferred 7 micron technology in 1976, has progressed to 90 nm line-width by 2005. In 1979, 6 years after ITRI was founded, the semiconductor maker UMC became the institute's first spin-off company. ITRI spun off another chip manufacturer, TSMC, 8 years later, in 1987, followed by the Taiwan Mask Company in 1988. Today, TSMC and UMC together have 70 percent of the world's IC foundry business. It was ITRI's role, Dr. Chu stated, "to continuously spin off and create new companies in Taiwan." Meanwhile, Taiwan's information industry, which posted $67.2 billion in 2004 revenues, has developed a wide variety of industries and products, led by such firms as Acer and BenQ.

Major Elements in Taiwan's Industrial Evolution

Behind Taiwan's industrial evolution, in ITRI's analysis, were four major elements: government policies; industrial infrastructure; foreign investment; and augmentation of technology. Dr. Chu explained each element in turn.

Government Policies

Dr. Chu related a timeline showing policies aimed at enhancing industrial development that Taiwan's government had implemented between 1950 and the present. Stressing certain points along the continuum, he said that in the early 1960s a duty-free export zone for manufacturing was established; around 1980, the Hsinchu Science Park was developed, followed in the late 1990s by the Southern Taiwan Science Park and the Southern Taiwan Innovative Park. Also reflecting government policy, a Food Industry Research and Development Institute was established as that industry developed in the early 1960s; ITRI's founding followed in 1973. The government had meanwhile built up basic infrastructure through 10 major public construction projects, and innovative programs were in store for the future as well.

Narrating how the government goes about allocating research resources in the service of technology development, Dr. Chu said the process begins with investment in infrastructure through the purchase either of a common facility and common equipment or of tools for strengthening existing facilities. Next, a promising new concept may be granted a small amount of seed money within the framework of the innovation plan of a university or other research institution. If it takes root there, additional project funding may be available under one of numerous programs dedicated to key technologies and components; although resources granted at this point may scale higher by an order of magnitude, the effort is still considered high-risk. If this further research is successful, the emerging technology is transferred to private industry in hopes that it will bear fruit through commercialization.

Industrial Infrastructure

Taiwan's industrial infrastructure is divided into four geographic zones. In the Northern zone are five cities, including Taipei; the country's international airport; and Hsinchu Science Park, with ITRI located nearby. It is about an hour's trip from Taipei to Hsinchu, which is 50 miles away. There are a number of universities and 38 incubators in the area. The zone has become a stronghold of ICT and IC sectors, owing to the industrial clusters there. Similarly, the other three zones have their transportation and education network, incubators, and industrial emphases.

Foreign Investment

Posting a table summarizing foreign investment coming into Taiwan in the latter half of the 20th century (Figure 17), Dr. Chu pointed out that the share of the total accounted for by the electronics and information-technology manufacturing sector, around 50 percent in the 1960s and 1970s, had fallen back to around 25 percent in the two subsequent decades. In contrast, the contribution of the

Unit: Percent

Year	Manufacturing				Services
	Food, Textile & Other	Chemical	Metals	Electronic & Information	
1950-59	23.98	47.20	0.46	9.59	18.00
1960-69	7.36	25.13	7.41	51.54	4.87
1970-79	6.65	16.11	20.54	47.57	9.12
1980-89	9.15	22.70	17.11	25.42	25.06
1990-99	5.01	10.28	11.42	28.08	44.60

Unit: Thousands of U.S. dollars

Year	Investment	
	Overseas Chinese	Foreign
1951-1959	9,305	10,874
1960-1969	123,980	276,175
1970-1979	608,817	1,223,286
1980-1989	991,563	7,705,845
1990-1999	2,025,577	23,983,142
Total	3,759,242	33,199,322

FIGURE 17 Foreign investments.

service industries had risen significantly since 1980, as the sector accounted for one-quarter of foreign investment in the 1980s and 44.6 percent in the 1990s.

Augmentation of Technology

There were three different paths to the augmentation of Taiwan's technology, Dr. Chu said: indigenous R&D programs, licensing and transfer from overseas companies or universities, and international cooperation. He displayed a graph (Figure 18) charting cases of international technical cooperation from 1950 through the end of the last century, which showed a peak in the 1980s. From decade to decade, it was Japan that had been Taiwan's leading partner, with the United States a distant second and Europe in third place. Over the previous 5 years, Taiwan developed numerous multinational R&D centers of various sorts and involving a wide variety of foreign companies.

Dr. Chu then explained the division of labor between Taiwanese academic institutions and industrial R&D organizations. The domain of the former, in addition to basic research, is the development of high-quality personnel to staff the country's research institutes and industry. The mission of the latter, of which there were more than 10, is not only to engage in technological innovation,

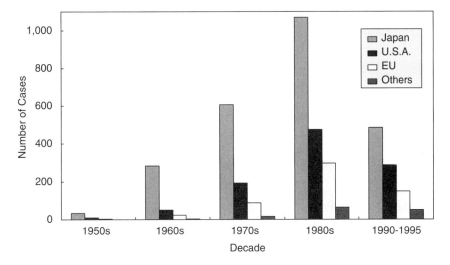

FIGURE 18 International technical cooperation.

development, and implementation, but also to facilitate the creation of new industries. Besides ITRI, which is the largest of these industrial R&D organizations, he named the Information Industry Institute, Food Industry Research Institute, Textile Industry Research Center, and Bicycle R&D Center from this category.

ITRI's Role in Taiwanese Growth

Focusing on ITRI, a not-for-profit R&D institute, Dr. Chu listed its tripartite mission: to create economic value through innovative technology and R&D; to spearhead the development of high-value industry in Taiwan; and to enhance the competitiveness of its industries in the global market.

ITRI has 13 research units—7 research labs and 6 research centers—divided into five areas: information and communications technologies; advanced manufacturing and systems; biomedical technology; nanotechnology, materials, and chemicals; and energy and environment. In 2004, the institute employed 6,540, of whom 14 percent had doctorates and just over 50 percent held master's degrees. Its goal for 2008 is to reach one Ph.D. for every five employees while maintaining the level of master's degree holders among employees at one-half. In addition to its current employees, ITRI has sent 17,000 alumni into Taiwan's workforce, 5,000 of whom are working at the Hsinchu Science Park, and this personnel transfer has been responsible for the creation of many new companies.

Of ITRI's $579 million budget in 2004, 52 percent came in the form of government R&D funding; the institute devoted between 20 and 25 percent of that to the development of high-risk technologies. Another 40 percent of ITRI's revenue came from technology transfer to industry, which the institute classifies as "knowledge-based services," while 6.6 percent was derived from its intellectual property.

ITRI's Hsinchu Chung Hsin Campus is a large complex housing its main headquarters, research laboratories, library, conference rooms, classrooms, dormitories, and dining, exercise, and medical facilities. Dr. Chu made special mention of the ITRI Incubation Center, known as the "Open Lab," which is also on the premises. The architecture is intended to promote integration, particularly among different areas of technology, as a tool for the development of new products and industries. In the previous 18 years, more than 100 new companies have originated there.

ITRI enjoys access to Taiwanese technological activity at every level: that of the universities, of other industrial R&D organizations, and of commercial enterprises. It engages in significant cooperation with universities and builds relationships to enterprises through technology transfer, international cooperation, human-resource development, and spin-off. As more than 90 percent of Taiwanese companies fall into the small and medium-sized categories, many lack the funding to do research independently, and ITRI helps them meet their needs for both knowledge-based services and research and technology development. ITRI links science parks, universities, and companies throughout Taiwan.

Cooperative arrangements connect ITRI not only with such major Taiwanese entities, but also with many important research institutions in the United States and Europe. As examples, Dr. Chu cited cooperation with the University of California at Berkeley focusing on nanotechnology and energy-technology development and with Carnegie Mellon University focusing on communications-technology research. He also mentioned ITRI links to Stanford University and SRI in the United States, the Netherlands' Organization for Applied Scientific Research, Russia's Ioffe Physico-Technical Institute, and Australia's Commonwealth Scientific and Industrial Research Organisation. Cooperating with MIT's Media Lab was ITRI's "Creativity Lab," established 3 years earlier to "link technology to the lifestyle" by focusing on the demand side in pursuit of new concepts in the consumer-applications sector.

Taiwan's Future Opportunities

Taiwanese policy makers see the country's future opportunities as grouped in three major areas—high-value advanced manufacturing, novel applications and products, and knowledge-based service industries—and believe that integration and innovation will be essential to taking advantage of them. To enhance its national innovation system in the era of the knowledge economy, Taiwan's government is pursuing the creation of basic infrastructure and an innovative business environment in order to strengthen relationships among the country's industry, academic institutions, and industrial R&D organizations. Some of the numerous industrial partnerships and alliances already in existence are the Taiwan TFT-LCD Association; Next-Generation Lighting Alliance; New Nylon & Polyester Textile R&D Alliance; Fresh Food Logistic Service Industrial Alliance; RFID System; Advanced Optical Storage Research Alliance; and Environmentally Friendly Manufacturing Technology Alliance. These entities' activities include resource deployment, standardization, patent pooling, market development, multidisciplinary integration, and coordinated development.

Still, Dr. Chu stated, Taiwan's "mindset and approach" are in need of adjustment in a variety of domains. When it comes to technology R&D, he said, the country was trying to move from optimization to exploration, from ordering work by single discipline to multidisciplinary integration, from conducting research in-house to collaborations and partnerships, and from developing components to developing system solutions. The value of a system solution was one to two orders of magnitude higher than that of a component, and the value of a comprehensive service system is higher still. Taiwan's intention is to proceed from components through systems to the service industries.

Summary and Conclusion

Recapping, Dr. Chu noted that Taiwan has moved from an agriculture-based society to an industrial economy in 50 years, creating many industries with significant global standing in the process. A key issue for Taiwan's continued success is close cooperation among industry, the government, and the academic and industrial R&D institutions. For the future, Taiwan will focus on new value creation through innovation to upgrade industry, while also trying to unify and align regional resources. The aim is to form productive clusters and facilitate the development of new service-sector industries by implementing innovative business models.

Dr. Chu invited those interested to seek information on ITRI at its Web site, *www.itri.org.tw*, and thanked the audience for its attention.

Discussion

Dr. Amsden opened the question period by inquiring about the criteria according to which firms are chosen to be admitted to Hsinchu Science Park. "How do you decide which firms to encourage?" she asked Dr. Chu. "Are you influenced by industry [or] simply by the quality of the firm?" She then invited a Swedish science reporter who was in the audience to step to the microphone. This reporter, asked Dr. Chu to describe further the innovative business model that he had recommended.

New Business Models, New Service Industries

Dr. Chu began by warning that the Taiwanese government's use of the term "service industry" may diverge from that of individuals and entities from other countries: It includes such activities as logistics, finance, and research consulting but not, for example, health care. One of the most important of ITRI's projects in the field of service-industry R&D concerns the innovative business model referred to by the questioner. In fact, after 30 years of technology development, ITRI's focus is to shift to developing new business models with the power to create new service industries.

As an example, Dr. Chu offered a logistical model that covers the manufacture of food from the delivery of raw materials through processing and on to distribution. New technology has emerged in the course of work on such value chains—in this instance, equipment was developed capable of keeping multiple temperatures constant for more than 24 hours—that in turn has led to the development of more new service-industry business models.

Selecting Technologies to Support

Kathy McTigue of the Advanced Technology Program, referring to Dr. Amsden's question about how firms are chosen for admission to Hsinchu Science Park, asked whether technology areas were chosen as well. She noted that Taiwanese innovation policy appeared to focus on promising technology areas. Research in these areas was supported through the Science Parks, and then transferred to private business. She asked if Taiwanese R&D is driven by the private sector (in the sense that private business initiates the projects and then comes to the government with requests for funding) or if the government itself chooses the domains for research. She also inquired about the relationship of ITRI to the Taiwanese government.

Dr. Chu answered that the government does provide private companies with funding to "encourage" them to embrace research projects chosen as the result of an evaluation process conducted by a committee. This "encouragement" usually takes the form of granting a company 25 percent of the research budget; the company is to put up 50 percent of the financing itself, with the remaining 25 percent covered by a government or bank loan. Similarly, when ITRI transfers technology it has developed to a company so that it can undertake product development, the government provides around 20 to 25 percent of the research budget in recognition of the risk involved; in these cases, a committee at the Ministry of Economic Affairs reviews the projects and decides where to place resources.

Concluding the presentation by Dr. Chu and the discussion that followed, Dr. Amsden introduced the next speaker, David Kahaner of the Asian Technology Information Program, to speak on Japan. Japan and its neighbors have had a very stormy relationship, she said, recalling that Japan had once held a great deal of East Asia under occupation. Yet when it comes to learning, technology transfer, and doing business in general, Japan's relationship with her neighbors is close. Many regard Japan as the hub of Asia's IT industry. Currently, Americans are looking very closely at Japan's innovation policies in an effort to learn from it. They are also asking how competition from China and other East Asian countries will affect Japan and, by implication, the United States. She then relinquished the podium to Dr. Kahaner.

JAPANESE TECHNOLOGY POLICY: EVOLUTION AND CURRENT INITIATIVES

David K. Kahaner
Asian Technology Information Program

Dr. Kahaner noted that he would begin by listing many of challenges confronting Japan today. Despite this list, he said, it should be "obvious to everyone that Japan is still a global technology powerhouse, a place where an incredible amount of extraordinary, world-class technology is produced and distributed." The prospect for Japan, therefore, is by no means entirely negative.

According to Dr. Kahaner, Japan's current challenges include:

- **An Anemic Economy.** Growth in the Japanese economy has been anemic for years. It is now showing some signs of turnaround, but the long-term trend is still unclear and the economy cannot be characterized as healthy.
- **An Aging and Shrinking Population.** Not only is the Japanese population aging, its growth is far from robust—if not, in fact, in decline.
- **Increased Global Competition.** Japanese firms face strong competition from low-wage countries like China as well as from more advanced countries like Korea, Taiwan, and Singapore. Japanese firms are engaged in outsourcing and insourcing alike, and some small manufacturers have brought their manufacturing back to Japan from lower-wage countries because they did not feel the quality they were getting abroad to be adequate.
- **A Less Favorable Business Climate.** Japan's business climate is not very good relative to that of other countries, as benchmarked in a variety of ways.
- **A Perception of Low Creativity.** A Western perception that Japan is suffering from a lack of creativity and from an associated lack of competition in its education system has, to a certain extent, "bled over to the Japanese themselves."
- **A Strong Currency.** The yen is appreciating against some currencies, making it more difficult for Japan to export.
- **Less Efficient Research.** R&D is not viewed as being very efficient.
- **Bureaucratic Obstacles.** Not only were there rivalries among Japan's ministries, there are also walls separating them.
- **Regulatory Burdens.** A large number of regulatory problems exist.

- **A Lack of Openness.** Japan is still viewed as a closed society by many in the United States.

Japanese Wrestle with the Role of Technology

The Japanese have long been wrestling with the questions of whether and how science and technology can help their country deal with some of these problems. Dr. Kahaner alluded to a variety of initiatives within Japan, ranging from more automation to the development of human capital from outside Japan, to new industry-academia-government collaboration. He called the preoccupation with creating knowledge-based industries "a kind of mantra" throughout East Asia, saying it could be seen in Korea, Singapore, and Taiwan as well as in Japan.

Japan's Science & Technology Basic Law

Beginning his discussion of Japanese policy, Dr. Kahaner evoked the promulgation in November 1995 of Japan's S&T Basic Law, which he described as an effort to "help the Japanese get their hands around where they were going in technology." Behind the law, he believed, is the goal not only of contributing to the country's economic development and social welfare by improving its technology, but also of contributing to the sustainable development of human society through the progress of S&T internationally. By the early 1990s, he explained, Japan was perceived by the United States and, perhaps, the countries of Europe as simply not pulling its weight in terms of international science and technology. "One could argue whether that's true or not," he said, "but the Japanese believed it," and they undertook to equalize their S&T investments with those of other countries relative to the size of the economy.

The Basic Law is one of the outgrowths of that effort, and it in turn resulted in the founding in January 2001 of Japan's Council for Science and Technology Policy (CSTP), which might be loosely compared to the United States' Office of Science and Technology Policy (OSTP). Chaired by the prime minister, the council is made up of six cabinet ministers, five academics, and two representatives of industry. It is, in effect, charged with developing the "grand design" for Japanese S&T policy; among other things, CSTP discusses new types of budget items, and its decisions influence each ministry's budget.

One of the council's most important duties is drafting the country's 5-year S&T Basic Plan, which sets guidelines for the comprehensive and systematic implementation of Japan's overall S&T promotion policy. The goal of the first Basic Plan, which went into effect in 1996 and thus predated CSTP's creation, was to double government spending on R&D. The second Basic Plan, whose budget was set at $212 billion over 5 years, was a part of an effort to double the amount available for competitive funding through the end of 2005. The main thrust of the third Basic Plan, to go into effect in 2006, was still under discussion at the time of the conference.

Allocating the Research Budget

Japan's total 2005 S&T budget was $36 billion, an amount 2.6 percent higher than in the previous year. Of that, around $13 billion was for research expenses including researchers' salaries, the remainder for infrastructure. In the budget's structure an umbrella function, called "systematization and integration" and denoted by an "S," overarched four key areas being developed for the future, each of which was denoted by a letter so that the whole formed the acronym "SMILE": nanotechnology and materials, "M," allotted 4.9 percent of resources; information technology, "I," 10.4 percent; life sciences, "L," 22.7 percent; and environment, "E," 7.5 percent. Again Dr. Kahaner pointed to a similarity between the Japanese approach and that of Taiwan as outlined by Dr. Chu.

Most important about the third Five-Year Plan, in his opinion, is that some form of aerospace-technology R&D is likely to take its place beside the four key research fields already mentioned. In addition, foreigners and women—the latter termed by Dr. Kahaner "the incredible untapped resource" of Japan—are to be sought out for a larger role in university research.

He pointed to a "new emphasis on efficiency" expected to be associated with a variety of policies, some involving collaboration. One locus of collaboration is to involve industry, academia, and government more actively—where, according to his personal perception, the Japanese saw the United Kingdom as more efficient than the United States. Another locus of collaboration, which is to receive greater emphasis in the third Basic Plan, is situated between the national government and local governments—once more, "quite consistent" with Dr. Chu's description of Taiwan.

Also of significance is a "big jump"—amounting to a 30-percent increase over 2004—in the money to be awarded through peer-reviewed competitions. To show the extent to which such competitions were already being employed, he displayed a table listing programs in the second Basic Plan's four key areas, with those granting funds on a competitive basis indicated in bold type (Figure 19).

Grant Competitions Heralding Change?

Dr. Kahaner suggested that the increase in funding granted competitively is in line with an emphasis on better integrating universities into the innovation system. Relatively few companies have spun out of the Japanese university system and relatively few patents have been produced by it—"a clear indication," he said, "that something was wrong." For this reason, the national universities have been converted over the previous few years into "independent administrative agencies," the rough equivalent of National Laboratories in the United States.

Program, **bold font means competitive research funding**	Ministry	2005 Budget in US$ Bil.
Life Science: **Molecular Imaging**	MEXT	1.150
Life Science: Research on Aids, Hepatitis, Emerging and Reemergi ng Infectious Disease	MHLW	4.526
Life Science: Efficient Breed Improvement Technology Based on Ge nome Breeding	MAFF	1.580
Information and Communication: Next Generation Back-bone	MIC	2.000
Information and Communication: **R&D for the Establishment of IT Infrastructure**	MEXT	2.974
Information and Communication: Human-assisting Robotics Realization Project	METI	0.900
Information and Communication: Autonomicus Movement Assisting Project	MLIT	0.490
Environment: **Promotion of Establishing Earth Observation System**	MEXT	1.017
Environment: Research on Agricultural, Forestal and Fishery Bio-cycle	MAFF	1.400
Environment: Development of Environmental Technology Based on Na notechnology	MEXT	0.400
Nanotechnology: **Fused Emerging Field Based on Nanotechnology and Materials**	MEXT	1.450
Nanotechnology: Nano-Medicine Healthcare	MHLW	1.416
Nanotechnology: Realization Project of Nanotechnology-Based Advanced Devices	METI	0.800

FIGURE 19 Major R&D Programs in 2005.

While still funded by the government, these agencies now have more auton-omy and more flexibility than traditional universities, exemplified by the ample opportunity they now have to seek competitive funds and now reflected in their increased cooperation with industry. Faculty and industry have in fact always cooperated, Dr. Kahaner said, but in a very informal manner; cooperation has recently become much more formalized. Laws have been enacted that allowed Japanese professors to "become millionaires if they're good enough and they have good enough ideas," he said.

There is also an effort to develop "Silicon Valleys" around the universities in keeping with the concept of local or regional clusters, as seen in Taiwan. In a development dating back only 5 years, each Japanese university has established a technology-licensing office; technology management as a discipline is also now receiving attention at a level unprecedented at Japanese universities.

Promotion of faculty based on performance has been adopted in some instances, although Dr. Kahaner said its full implementation in Japan "would be an incredible thing." More non-Japanese faculty are receiving regular appoint-ments, in contrast to the traditional practice of taking on foreigners almost exclu-sively under one-year contracts. An "overt goal" to make 30 Japanese universities world-class according to objective international criteria has brought about both alliances and consolidations among institutions. The country, he noted in passing, has around half as many colleges and universities as the United States, so that on a per capita basis the two nations were on an equal footing.

Ministry	Agency	Research Orientation	Recipients
MEXT	JSPS	Basic research	Universities
	JST CREST PRESTO ERATO ICORP	Mission-oriented basic research targeted for commercialization in 10-20 years	Universities & National labs
METI	NEDO	Application-oriented R&D targeted for commercialization in 5-10 years	Industry, Universities, National labs collaborations
MIC	TAO	Specialized in telecommunications	Universities/National labs

FIGURE 20 Funding agencies and missions in Japan.

Administering S&T at the Cabinet Level

Returning to the subject of interministerial rivalry, Dr. Kahaner noted that in 2001 essentially all Japanese ministries were reorganized. For example:

• The former Education Ministry and Science & Technology Agency are now merged into the new Ministry of Education, Science, Culture, Sports, and Science & Technology (MEXT).
• A Ministry of Public Management, Home Affairs, Posts and Telecommunications was renamed in 2004 to Ministry of Internal Affairs and Communications (MIC).
• The Ministry of International Trade & Industry (MITI) is now reborn as the Ministry of Economy, Trade & Industry (METI).

Most significant about these mergers is that they have made Japan's central government smaller. This has strengthened Japan's Cabinet Office, which in turn both lends more weight to the decisions of organizations like the Council for Science and Technology Policy and puts more power of coordination into their hands. See Figure 20 for a summary.

Under the most recent Basic Plan, Japan's ministries are being encouraged to collaborate on work taking place within eight specific R&D groupings. These

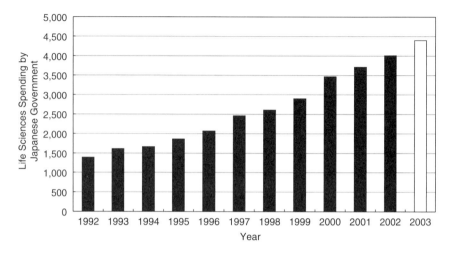

FIGURE 21 Life sciences—biotech: Japanese government spending in life sciences.

groupings include post-genomic research, hydrogen utilization and fuel cells, nano and bio technologies, and ubiquitous networks. Recent research thrusts for METI are fuel cells, robotics, health, IT in the home, energy, and nanotechnology, the last three of which would, Dr. Kahaner predicted, be crucial in Taiwan and Korea as well. In another recent development, Japan has begun taking much the same approach as Taiwan in nurturing knowledge clusters by emphasizing specific fields of technology; CSTP is establishing more than a dozen of these clusters around the country over the previous several years.

Japan's Leading Endeavors in Science

Dr. Kahaner then turned to what he termed Japan's "big-time science efforts":

Biotechnology

According to a graph he projected, the government tripled its funding for the life sciences in the decade beginning in 1992 (Figure 21), with the goal of increasing the number of biotech companies in the country to 1,000 by 2010. The number of Japanese biotech companies has grown from 60 in the late 1990s to 250 in September 2001.

Country/ Region	Population in millions	GDP in trillions of U.S. dollars	Growth (percent)	Per Capita GDP in thousands of U.S. dollars	Research as a percent of GDP	Nano Budget 2004 in millions of U.S. dollars
U.S.A.	293	11	3.1	37,500	2.8	**961**
E.U.	456	11.1	1	24,300	1.9	~ 800
China	1,300	6.45	9.1	5,000	~ 1.0	~ 50
Japan	127	3.6	2.7	28,300	2.9	**~ 940**
India	1,065	3.03	8.3	2,800	~ 0.5	~ 25
Korea	48.6	0.86	3.1	18,000	2.7	248
Taiwan	23	0.53	3.2	23,000	2.3	90
Singapore	4.4	0.11	1	23,700	2.2	~ 12

FIGURE 22 Spending on nanotechnology.

Nanotechnology

Japan is spending almost as much in nanotechnology ($940 million) as the United States ($961 million) on an absolute basis in 2004 (Figure 22), and much of the funding is granted competitively.

Fuel cells

All Japanese electronics companies are working in this area, which is a focus of government funding because of the ever-growing use of portable devices. Dr. Kahaner pointed to an announcement, made only a few days before the conference, that IBM and Sanyo were to develop a fuel-cell power system for the new-generation Think Pad; it was his understanding that most of the basic technology development would take place at Sanyo.

Robotics

Unlike in the United States, there is a very high level of interest in Asia—and particularly in Japan—in humanoid robots, the first of which, Honda's Osimo, had come out in the late 1990s. Of interest to the Japanese, Dr. Kahaner suggested, is not so much the robot as the development of capabilities associated with its attributes: power-source technology; recognition technology, including voice, tactile, vision, and translation; activator technology, pertaining to the robot's mechanics; structure, of possible relevance for prosthesis technology;

control-system technology; and software. "Think about all of this as component technology, and about the impact that developments in these technologies can have across the industry spectrum," he advised. "Don't think of it in terms of this walking robot."

Synchrotron radiation

SPring-8 was the largest synchrotron radiation facility in the world when it opened in 1997 and is still among the biggest. Its $1 billion price tag is comparable to that of the U.S. Department of Energy's Spallation Neutron Source at Oak Ridge, Tenn.

Computing

Japan's Earth Simulator, the world's fastest scientific computer when it opened in 2003, required a $450 million investment and "still sends shudders through many people in the U.S. science community," Dr Kahaner said.

Concluding, he said that Japan will place emphasis in the future on competition within the country for research money; collaboration among Japanese organizations and the associated coordination of research efforts; ways of being more efficient and of measuring efficiency; and increasing internationalization with respect to human talent.[12]

Discussion

Opening the question period, Jim Mallos of Heliakon observed that the point of all innovation was to make new products and that, in the 21st century, "it won't do us any good to come up with new products that are like the failed competitors of the iPod: perfectly good, but not quite magical in the way they put together design and technology." His question was whether Japan's esthetic culture and skill at miniaturization positions it as well as any nation to succeed at modern product design.

Dr. Kahaner said that the answer is "obviously" affirmative and that the Japanese themselves believe that they were very well positioned in this regard. In not only Japan but also Taiwan a vast array of products that differ only slightly among one another are "being pushed out the door in very, very great numbers [as] a way of experimenting with what the public finds suitable." While such products got "filtered out" before they reach the United States market, he remarked, they are greatly in evidence in Akihabara and similar places throughout Asia. So it was indeed likely that Japan will be a strong competitor.

---

[12]See the related presentation on Supercomputing by Kenneth Flamm in Panel IV of these proceedings.

On the specific question of whether the Japanese would produce "the next-generation iPod," however, Dr. Kahaner declined to guess. He stated his personal view that devices like the iPod would lose out in the future to a telephone-based technology. Noting that there were hundreds of millions of cell phones in use but "only a handful of iPods," he said that market pull would best be achieved by starting with the cell phone and innovating from there rather than by innovating on the basis of the iPod.

Dr. Amsden then called on the next speaker, Tom Howell, to talk on China's semiconductor industry.

NEW PARADIGMS FOR PARTNERSHIPS: CHINA GROWS A SEMICONDUCTOR INDUSTRY

Thomas R. Howell
Dewey Ballantine

Referring to Dr. Amsden's observation that in developing countries industry comes first and innovation later, Mr. Howell said he would allow that progression to structure his presentation.

Over the previous 5 years, China's technology level and the scale of Chinese industry has grown on a very fast trajectory. In 2000, conventional wisdom had it that China would lag behind the world leaders in semiconductors for quite a long time; according to the current conventional wisdom, it has become an "unstoppable juggernaut." In Mr. Howell's view the conventional wisdom is most likely wrong in both cases, but he acknowledged that the speed of the country's development has surprised almost everyone. In the course of his talk, he was going to examine some of the government policies that have contributed to it.

Overcoming a Legacy of Obstacles

A snapshot of China's microelectronics industry taken in the early 1990s would have shown that the country had destroyed its science infrastructure during the Cultural Revolution. The universities had been closed, many people had been driven out, and an entire generation was still feeling the effects of not having received an education. All Chinese semiconductor enterprises and research organizations were owned by the government and administered by bureaucrats. They were 10 to 15 years behind the global state of the art; at that time this was attributed to what was called the "Western technology embargo," but the fact was that Chinese firms were very far behind technologically.

Because the industry's infrastructure was primitive, Western companies had very little interest in building facilities in China. The level of excellence in Chinese semiconductor manufacturing was reflected in a comment by a Texas Instruments employee: "I've seen clean rooms with open windows." Virtually all of the integrated circuits imported into China were smuggled through "various shady trading companies" in Hong Kong. To the extent that foreign investors were invited in, they were subjected to crude pressure from the government to transfer technology. In sum, "it was not a very welcoming environment for foreign investment."

One element of the Chinese effort, however, has been a continual study of successful systems abroad, particularly those in the United States, accompanied by constant critical self-appraisal. The analysis of what China was doing wrong was applied over and over again, Mr. Howell said, and "managed to trump all the other disadvantages eventually." This process is still going on.

The Structure of China's Development Policy

Projecting a table listing domestic policies and practices that have affected China's development in microelectronics (Figure 23), Mr. Howell characterized as "all stick, no carrot" the traditional Communist policies current in 1994 (center column). "You have state-owned enterprises, and the government tells them what they're supposed to do, and they're expected to do it," he summarized. Foreign investment was restricted to the point that China "was essentially a closed market," and tariffs on semiconductors as well as many other electronics products were high. There were about 100 "high-tech parks," none of them very sophisticated.

But while China, with its billion people, was lagging technologically, nearby Taiwan—a resource-poor Chinese society of 20 million people—was building one of the foremost semiconductor industries in the world. "That fact was not lost on planners across the Fujian Straits," Mr. Howell remarked. "As they were doing their self-criticism, one could point out that 'we're doing things wrong that the Taiwanese are doing right; perhaps we could learn something from them.' "

Recalling Dr. Chu's summary of Taiwanese policies and referring to his own chart (right-hand column), Mr. Howell said that Taiwan's government basically functioned (and continues to function) as a partner with industry. Rather than making decisions it encourages the formation of enterprises, in some cases spinning them off from government research institutes. A passive equity investor in many enterprises, the Taiwanese government becomes involved when it is needed to create infrastructure or where there is some task or risk that the private sector can not undertake on its own, such as the pioneering of the pure-play foundry. Intervening at such points but not at others, Taiwan has presided over "an all-carrot system" in contrast to the previous Chinese system, which was "all stick."

China Adopts a New Policy System

Around the beginning of its tenth Five-Year Plan in 2001, and concurrent with joining the World Trade Organization (WTO), China undertook a funda-

Policy/practice	China 1994: Command Economy Mode	Taiwan 2000: Partnership Model
Principal form of leading semiconductor enterprises	State-owned enterprise	Private, gov't holds passive minority share
Business model of leading semiconductor firms	Integrated device maker	Foundry
Policy toward foreign direct investment	Heavily restricted	Liberalized
Promotion of IC design industry	Emphasis on state-owned research institutes	Privatization of gov't research institutes Financial assistance to private companies
Government as direct investor in leading firms	100% gov't ownership of semiconductor enterprises	Government passive minority equity stake
Tariffs on semiconductors	6-30 percent	0
Industrial parks	Over 100 "Hi-Tech parks"	1 flagship park (Hsinchu), 2-3 others emerging (Tainan, Nankang)
Major financial incentives to individuals	None	Major tax benefits
Government controls enterprise decision making	Yes	No
Government promotion of venture capital sector	No	Yes

FIGURE 23 China 1994, Taiwan 2000.

mental reappraisal of what it was doing and "essentially decided to jettison [its] whole system." While retaining the economic nationalism that had suffused all its earlier Five-Year Plans, it largely abandoned the command method in favor of a system using Western promotional measures permitted under the WTO: subsidies, tax measures, targeted government procurement, and the like. Simultaneously taking place was a thorough decentralization, with most of the policies being implemented locally rather than at the national level; a fundamental redefinition of the industry-government relationship, with an emphasis on the independence of enterprises' decision making; and liberalization of inward investment permitting foreign companies to establish fully owned subsidiaries. Tariffs on semiconductors were eliminated. And pressure to transfer technology eased, although that pressure had not ceased entirely.

This added up to a "paradigm shift" in which Chinese planners abandoned their own system and embraced Taiwan's. Mr. Howell displayed a table showing that virtually every Chinese policy current in the semiconductor field had a Taiwanese antecedent (Figure 24).

Policy/Practice	China 1994: Command Economy Mode	Taiwan 2000: Partnership Model	China 2002
Principal form of leading semiconductor enterprises	State-owned enterprise	Private, government holds passive minority share	Private, government holds passive minority share
Business model of leading semiconductor firms	Integrated device maker	Foundry	Foundry
Policy toward foreign direct investment	Heavily restricted	Liberalized	Liberalized
Promotion of IC design industry	Emphasis on state-owned research institutes	Privatization of government research institutes Financial assistance to private companies	Privatization of government research institutes Financial assistance to private companies
Government as direct investor in leading firms	100% government ownership of semiconductor enterprises	Government passive minority equity stake	Government passive minority equity stake
Tariffs on semiconductors	6-30 %	0	0
Industrial parks	Over 100 "high-tech parks"	1 flagship park (Hsinchu), 2-3 others emerging (Tainan, Nankang)	1 flagship park (Zhangjiang), 2-3 others emerging (Suzhou, Beijing)
Major financial incentives to individuals	None	Major tax benefits	Major tax benefits
Government controls enterprise decision making	Yes	No	No
Government promotion of venture capital sector	No	Yes	Yes

FIGURE 24 Microelectronics: China embraces Taiwan's model.

Many of them were implemented with the assistance of Taiwanese advisers; for example, one of the leaders who had set up Hsinchu Park, Irving Ho, acted as a consultant on the industrial parks that had been built on the Mainland in the previous 5 years. The function of China's central government in policy had become "mostly hortatory," with the actual benefits and promotional measures implemented largely by the regional governments and local governments in line with the central government's intentions (Figure 25).

China's Market Pull Restructures the Industry

A policy not copied from Taiwan was the leveraging of China's market. When Mr. Howell spoke in 2000 with government and industry planners in Taiwan, their expectation was that 30 new semiconductor fabrication plants would be built in the Mainland by 2008 and that there would be no Taiwanese investment in semiconductor manufacturing on the Mainland. Two years later, the picture had changed radically: Seven Taiwan-invested fabs were envisioned for Taiwan, 20 for China. Two of the Mainland fabs were already operational in 2005, the rest either under construction or planned by companies that were managed and, in many cases, capitalized by Taiwanese entities.

Measure	Central government	Regional government	Local government
VAT	• Provides 3-6% VAT preference	• Delegates authority to associations to determine enterprise eligibility	• May rebate local portion of VAT to enterprise • May apply preferences more broadly
Enterprise tax	• Provides tax holiday	• Provides exemption from local taxes/fees	• May rebate local portion of tax
Individual tax			• Special tax rates in parks • Rebates to individuals for housing/cars
Equity investment	• Ownership of SOEs, laboratories	• Equity participation in semiconductor enterprises	• Equity participation in semiconductor enterprises
Interest rate subsidy	• One point	• One-two points	
Land use fee			• Provides preferential terms (including free land)
Utilities			• Provides preferential terms
Infrastructure construction	• Encourages infrastructure projects • Approves, sets guidelines for new high-tech parks	• Provides funding, land, guidance for infrastructure projects, including new high-tech parks	• Constructs sites providing infrastructure for semiconductor industry
Attracting overseas talent	• Sets broad policies	• Grant residency • Provide venture capital, incubators for returnees	• Provide incubators • Offer individual incentives (cash, facilities, housing, car)

FIGURE 25 Layering of Chinese promotional policies in microelectronics.

Although this shift had been attributed by some to an advantage in manufacturing costs that China is assumed to enjoy over Taiwan, these costs—not counting government incentives—are in fact very similar, even close to identical. Richard Chang of the Semiconductor Manufacturing International Corporation (SMIC), one of the foundries on the Mainland, places the differential at less than 10 percent and probably closer to 5 percent. Cost considerations are not behind this locational shift in Taiwanese investment, Mr. Howell asserted.

Incentives Create a Cost Differential

Very strong, however, is the draw of the Chinese market, and China's government has emphasized its market's pull by putting a differential value-added tax (VAT) into effect in 2000 that gave devices manufactured in domestic fabs a 14-percent advantage over imports in the Chinese market. Taiwanese investors, seeing that they might be shut out of China's growing market unless they invested there, rushed across the Strait as a result. Although the VAT measure was subsequently withdrawn, that Mainland investment has remained—providing an example of China's use of its market to leverage inward foreign investment.

That market is expanding faster than any other of the world's major markets: It has grown at a rate of 40 percent in 2004 and is expected to achieve a compound annual growth rate of more than 20 percent for the period 2002-2008, compared to 7.3 percent for the United States and 13.8 percent for Taiwan. Mr. Howell put its current size at around $24 billion and said it was expected to grow to something on the order of $65 billion by 2007.

The bulk of the country's capital stock in semiconductor manufacturing still falls within the category of technologies that are "essentially obsolete." Yet, while 8-inch fabs are predominantly being built, there is also a 12-inch fab located in Beijing that had gone into operation in the second quarter of 2005, and still more growth is to be expected at the high end of the scale.

A "Leap Forward" for Chinese Technology

According to the U.S. General Accounting Office (renamed the General Accountability Office in 2004), China closed the wide technology gap in IC line-width capability that had existed in 1986 and by 2001 approached the U.S. state of the art.[13] The most advanced SMIC fab, in Beijing, is currently using design rules of 110 nm. While this is "not state of the art," Mr. Howell said, it is "not that far behind," as much of the U.S. industry is at 90 nm and moving to 65 nm. China has thus made "quite a leap forward."

Mr. Howell observed that while the old-style, 100-percent government-owned companies Huajing and Hua Yue are still in existence, they are "not doing much that is of much interest." China first moved to 50-50 joint ventures that it still essentially controls, but are partially capitalized by (mostly) Japanese investors; while "not failures," these ventures did not "chase the state of the art" either. However, the new-model China-based semiconductor ventures more closely resemble a typical multinational corporation with some government investment, but basically backed by a diverse array of stockholders, including foreign investors.

Pulling in Foreign Capital

Commenting next on the financial structure of SMIC, currently the largest foundry operator in China, Howell noted that, as of 2002, Chinese banks helped get SMIC up and running with close to $500 million in loans. The Industrial and Commercial Bank of China, China Construction Bank, Shanghai Pudong Development Bank, and Bank of Communication are policy banks and lend for policy reasons; the lenders were simply commercial banks. Contributing equity

---

[13]U.S. General Accounting Office, *Export Controls: Rapid Advances in China's Semiconductor Industry Underscore Need for Fundamental U.S. Policy Review,* GAO-020620, Washington, D.C.: U.S. General Accounting Office, April 2002.

along with the PRC Government and Shanghai Industrial Holdings (the investment division of the Shanghai municipal government) were the government of
Singapore, Toshiba, Chartered Semiconductor, and a number of offshore venture-
capital companies, many of them with Taiwanese-sourced money. SMIC was subsequently listed on the New York Stock Exchange in 2004; Mr. Howell estimated
its current equity ownership at 58 percent American.

As for the tax rates applying to enterprises operating in China, he said "there
is essentially no tax." Allowing companies to exist in a tax-free environment was
a duplication of Taiwanese policy.

Modernization of China's High-Tech Parks

Mr. Howell then traced the evolution of high-technology parks in China.
"In the old days," he recalled, "if you went into a [Chinese] high-tech park, you
wouldn't even know you were there: You'd just be in some kind of ramshackle
urban environment that had been designated as a park and was run . . . by an
administrative committee appointed by the municipal government." Taking a
first step away from this, the Chinese began setting up development corporations
that negotiated with investors and tried to start their own businesses. "But if you
went in to meet with them," he said, "you'd find that the person who met with
you had two business cards: an administrative committee card and a development
corporation card." Because these officials wore both hats, it was helpful as part of
negotiating a deal with a corporation to get a permit approved by the administrative committee, "because it was the same guy." Many Chinese parks are still run
in this manner, although it is not conducive to a very dynamic, entrepreneurial
environment.

The latest-model high-tech park is a sophisticated development corporation
with a venture-capital arm, a real estate development corporation that sells land in
the park, high-tech incubators, and many other support functions. The administrative committee manages the park's utilities and services. The Beijing Technical
Development Area is an example of this genre, resembling what one might find
in California. Mr. Howell noted that similar parks are located at Shanghai and
Suzhou. They amount to clusters of relatively advanced semiconductor manufacturing facilities that incorporate all the necessary infrastructure: materials companies, equipment companies, design centers, and the design houses of various
OEMs. There is also plenty of land.

China Aims to Innovate in Microprocessors

Finally, Mr. Howell said that, having been successful in drawing industry
in, the Chinese are now seeking "very vigorously" to develop their domestic
innovative capability. Chinese officials have been making statements in recent
years that Intel has a relative monopoly in semiconductors, that Microsoft has a

monopoly in software, and that both monopolies are bad. In addition, Chinese officials have said that their response to the U.S. hardware monopoly is to try to design an indigenous microprocessor and have set in motion five projects whose goal it is to accomplish this capability. One of these projects is run by a government institute that is part of the Chinese Academy of Sciences—The Institute for Computer Technology—that in 2002 spun off a microprocessor company. He rated its product as "pretty good—not as good as a Pentium, obviously, but they're closing the gap."

The country's end-use industries have been organized as well and are being encouraged by the government to buy the output through the "Dragon Chip Industrialization Alliance," with their purchases to be complemented by military procurement. Since China did not sign the WTO procurement code when it joined the organization, the Chinese can practice discrimination in procurement when doing so conforms to their interest. "They feel now that buying systems that have Chinese-developed microprocessors in them is not in their interest," he observed. "But the minute that the Chinese model gets to be as fast as an Intel model, they will all switch at one time to the Chinese model." Considering the current and projected growth of the Chinese market, such an occurrence could be expected to have "quite an impact."

Assessing the Challenges Facing China

In conclusion, Mr. Howell presented a list of "concerns and challenges":

- He all but dismissed oft-expressed fears of **overcapacity**, suggesting that China's government could take such demand-boosting measures as requiring each citizen to have an I.D. card with an embedded microprocessor if it needed to absorb excess output as the number of fabs in the country grew.
- The use of **preferential government procurement** as an industrial policy tool is, he said, a concern.
- **Intellectual property rights**, although growing, are not yet a major concern, because China's technological level has not reached that of the United States.
- **Standards setting** is a concern because it can be used to shut out American and other foreign designs.
- **Government pressure to transfer technology**, although more subtle than before, has not gone away.

Mr. Howell rated as the biggest challenge (what he called) the "gravitational pull" that increasingly draws all levels of semiconductor industry activity to China. As the bulk of wafer-fab investment moves to China—and projections indicate that China will boast some 30 new fabs in the ensuing 3 years compared to 6 new fabs in the United States—science and engineering graduates from universities around the world would increasingly find the opportunities they were seeking in China. Combined with the growing location of design capacity in China, "a tipping point has been reached that we can't easily turn around," he warned.

Discussion

While agreeing with Mr. Howell about the importance of foreign-owned firms to China's semiconductor industry, Dr. Amsden cautioned that it would create the wrong impression to say that foreign firms are important in all Chinese industries. One of the Chinese government's major policies is to create nationally owned firms, private or not, in virtually every major industry, including semiconductors. "The idea is that you have joint ventures," she said, "but unlike in other countries they have a finite life span: After 10 years they're dissolved." China's emphasis on nationally owned firms as opposed to relying on multinational firms is an extremely important element in its development, a subject on which, she suggested, Mr. Howell might comment later on.

Then, informing the audience of their good fortune at having him present, she ceded the rostrum to Mr. Shindo.

INNOVATION POLICIES IN JAPAN

Hideo Shindo
New Energy and Industrial Technology
Development Organization (NEDO)
Japan

Mr. Shindo expressed his pleasure at having the opportunity to introduce some of Japan's activities in the domain of innovation policy at the symposium.

He began by projecting a diagram illustrating the nature of the relationship between NEDO and METI (Figure 26), the former being a funding agency closely connected to the latter, as Dr. Kahaner had explained earlier. He reported that the term "innovation" is very popular in Japan and is regarded as highly important. Japan's Basic Science and Technology Plan has addressed innovation from a science and technology perspective. In addition, the *Nakagawa Report: Toward a Sustainable and Competitive Industrial Structure* has addressed innovation from the viewpoint of industrial policy. Published in 2004, this report was named after Shoichi Nakagawa, Japan's Minister of Economy, Trade & Industry.

Mr. Shindo focused on two key questions, one centering on invention, the other on innovation. The first concerns how new ideas are created and casts the spotlight on how to provide a good environment for R&D, whether at the basic stage or on technologies progressing through the Valley of Death. The second concerns how to introduce such ideas to the market. It focuses on what he called

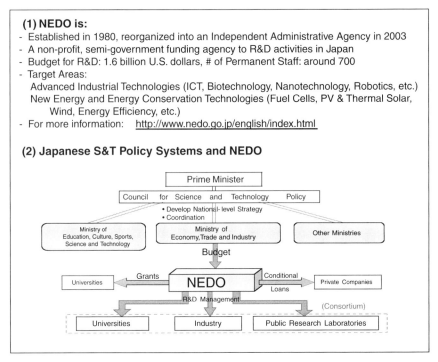

(1) NEDO is:
- Established in 1980, reorganized into an Independent Administrative Agency in 2003
- A non-profit, semi-government funding agency to R&D activities in Japan
- Budget for R&D: 1.6 billion U.S. dollars, # of Permanent Staff: around 700
- Target Areas:
  Advanced Industrial Technologies (ICT, Biotechnology, Nanotechnology, Robotics, etc.)
  New Energy and Energy Conservation Technologies (Fuel Cells, PV & Thermal Solar,
     Wind, Energy Efficiency, etc.)
- For more information: http://www.nedo.go.jp/english/index.html

(2) Japanese S&T Policy Systems and NEDO

FIGURE 26 What is NEDO in the S&T policy system in Japan?

"the virtuous cycle of demand and innovation." He interpreted the "innovation ecology" referred to by Secretary Marburger as a way of thinking about the cycle involving innovation and the market in order to enable more sustainable growth for innovative technologies.

Key Questions for Japan's Next Five-Year Plan

After briefly providing background information on Japan's Basic Science and Technology Plan, Mr. Shindo highlighted a number of possible key questions regarding the third Basic Plan:

- How to develop and maintain S&T human resources?
- How to establish a creative, high-quality R&D system?
- How to prioritize strategic S&T areas?
- How to develop a virtuous cycle of knowledge creation and application, through the acceleration of innovation and value creation? This virtuous cycle would be based on R&D results at universities and public research organizations

on the one hand, and through industry-university cooperation, the activation of entrepreneurs, and human resource development in management of technology areas on the other.

The *Nakagawa Report*, concerned above all with the activities of METI, seeks to identify policies needed to establish and accelerate a virtuous cycle of demand and innovation in order to bring about Japan's economic recovery and to create its future industrial structure. Taking a "very traditional" approach to drafting this report METI staff conducted rigorous interviews with over 700 people from more than 300 companies and institutions, asking all what they felt to be important.

METI Looks at Japan's Future

The report provides three key questions and as many key solutions. The questions center on how to ensure global competitiveness, how to respond to the demands of society, and how to encourage regional economic development. The potential solutions are to identify cutting-edge areas of industry that promise strong global competitiveness industrial areas that can meet market needs arising from changes to society, and industry clusters that can support regional revival.

Also contained in the report is a "very comprehensive" list of policy priorities. The first identifies promising industrial areas, among which, as mentioned earlier by Dr. Kahaner, are fuel cells and digital consumer electronics. Second are policies for regional revitalization. A third category includes so-called cross-sectional policies, pertaining to such issues as the development of industrial human resources, intellectual property rights, research and development, standardization, development of new businesses by small and mid-sized enterprises.

Turning to the implementation of these potential solutions and policies, Mr. Shindo said that "fortunately" several policy responses to the report have been initiated in the year since its publication. Notably, a "Technology Strategy Map" that has been developed by his organization together with METI and the National Institute of Advanced Industrial Science and Technology (AIST), Japan's largest public research organization.

Ending his presentation, he thanked those attending for their attention and offered to provide more details to all who might be interested.

DISCUSSION

Dr. Wessner opened the question period by referring to the "tipping point" mentioned by Mr. Howell and asking whose semiconductor industry was most vulnerable to it—that of the United States, Europe, Taiwan, or Japan? While he was posing the question primarily to Mr. Howell, Dr. Wessner said, he would value the response of others on the panel as well.

Mr. Howell responded that, in his opinion, Taiwan's semiconductor industry is the most vulnerable. A "huge exodus" from Taiwan to the Mainland has already taken place, and hundreds of thousands of Taiwanese are how living in China, especially in the Shanghai area. Companies are springing up there that are run by Taiwanese managers and staffed in the main by Taiwanese engineers. The Chinese have figured out that even if companies can not be lured to the Mainland, individuals can be lured there and brought together, along with investment and other elements needed to create companies. "It's a very attractive environment if you're Taiwanese," he commented. "You speak the language; and Shanghai is pretty nice, really; and opportunities are there for many Taiwanese that they may not see for themselves in the long run in Taiwan." What Taiwan's strategy was for responding might be, he said, has not yet become clear.

Comparing Costs in Taiwan, on the Mainland

In answer to a question about the extent to which cost levels differed between Taiwan and the Mainland, Mr. Howell said that labor costs for semiconductor manufacturing are about 40 to 50 percent lower on the Mainland and that the cost of water is lower as well. But these are very small components of semiconductor manufacturing costs; the main expenses are equipment and other items whose costs are equivalent or close to equivalent in the two locations and were in some cases lower in Taiwan.

In addition, the Taiwanese fabs on the Mainland have the extra costs associated with bringing in expatriates to work. Housing and education might need to be provided to hundreds of people coming from Taiwan, while those from Europe, the United States, and Japan have "expectations of a certain lifestyle," and catering to these expectations raise labor costs. A 2002 Dewey Ballantine survey comparing costs in the United States, Taiwan, and China found manufacturing costs to be very similar in all three locations: "a little bit lower in China and little bit higher in the United States, but not that much if you take incentives out of the picture."

Adding incentives, however, changes everything, as demonstrated by China's erstwhile VAT policy, which pushed cost significantly lower. It was thus an artificial cost advantage created by the Chinese government that motivated companies to move from Taiwan to the Mainland. "Right away you got a 13-percent cost advantage just based on that," Mr. Howell pointed out. "And there was a sense that you would have the market to yourself if you moved over there with advanced technology: Nobody could export into that market and meet that cost advantage. It sucked in just an incredible amount of investment, and people and skills and everything else."

Evolution of Technology Partnerships in the United States

Moderator:
Lewis S. Edelheit
General Electric, retired

Dr. Edelheit, GE's chief technology officer during the 1990s, opened the session with some observations from what he described as an industry perspective. "This symposium," he said, "is about competitiveness: Some countries are trying to figure out how to get it, others how to keep it, and still others how to get it back. And it's all about learning how to move fast and win in a brutally competitive economy such as we've never seen."

About a quarter-century before, he recalled, industrial laboratories in the United States had become "very uncompetitive." No longer relevant to their businesses, they were in a position of having to change, and change a great deal, or else die; meanwhile, the new companies that arose were not forming research labs. The deciding factor was not so much the issue of applied versus basic research, or even the issue of short-term versus long-term research, but the fact that the old model just was not producing results. "Funding a basic researcher in a lab someplace to do something, then hoping that you could get it into production," he said, "stopped working. It was too slow."

Many industrial research labs did die, but others changed, and very quickly. Of the numerous ways in which they changed, Dr. Edelheit named two. One was to start partnering with their own businesses much more closely, as the only way they could acquire speed was to move much closer to the marketplace. The second was to start partnering with other companies, with other industries, with government, and with other countries, since even a company as big as GE was unable to move rapidly enough on its own.

UNITED STATES USING OUTDATED RESEARCH MODELS

But, in many cases, the old models remain in use in the United States—within the government, industry, universities, and the National Laboratories—and they were not moving fast enough. Other countries are wrestling with how to speed up the innovation process, especially in such areas as energy, job creation, health care, and the environment, where national needs are not being met with sufficient speed, in part because the models are too slow.

It was to consider these questions that the symposium was convened, Dr. Edelheit said, and the current panel would offer three speakers with interesting perspectives on the issue of partnerships or the lack thereof, among government, industry, and universities in the United States. The first, Ken Flamm, would talk about the case of supercomputers, a technology that clearly was critically important for a great number of industries.

U.S. POLICY FOR A KEY SECTOR:
THE CASE OF SUPERCOMPUTERS

Kenneth Flamm
University of Texas at Austin

Dr. Flamm, thanking Dr. Wessner for the invitation to speak, said he would use his time to "tell a story" about the supercomputer industry. Some of the material he would present has already been published in an earlier version, having formed the basis for recommendations of a 2004 report on the future of supercomputing by a National Academies panel on which he served.[14]

He would begin with the field's early history, he said, in order to make sure that his listeners understood what is meant by the term "supercomputer" and where the supercomputer came from. The machines built to decrypt code traffic during World War II were the direct precursors of the modern electronic digital computer and the famous ENIAC machine, built for other purposes, appeared shortly thereafter.

In the Beginning, All Computers Were "Super"

At the computer industry's very beginning in the 1940s and 1950s, all computers were essentially supercomputers—every new model being the "biggest,

---

[14]National Research Council, *Getting Up to Speed: The Future of Superconducting*, Susan L. Graham, Marc Snir and Cynthia A. Patterson, eds., Washington, D.C.: The National Academies Press, 2005.

baddest, greatest computer ever built"—and a national-security application, whether bomb design or cryptography, was behind the funding of most computer R&D. It was in the late 1950s that a commercial computer market began to develop, and the term "supercomputer" came into use in the early 1960s. It was probably first applied to the IBM 7030 stretch, a special-purpose machine designed for two powerful government-mission customers, the National Security Agency and the Department of Energy. The Control Data 6600 was the other model referred to by that label, but in reality predecessors of both had merited it.

While all computers were supercomputers originally, as the commercial market began to develop and differentiate in the early 1950s, the gap separating the capability of the most powerful computer being produced and sold from that of the least powerful approached an order of magnitude. That gap grew to between three and four orders of magnitude by the end of the decade and, by the early 1970s, came to exceed the upper end of that range.

Control Data and Cray Constitute the Industry

From about the mid-1960s to the late 1970s, the entire supercomputer industry basically resided in two U.S. firms, Control Data and Cray, the latter company being a spin-off staffed by ex-employees of the former company and of its predecessor, Engineering Research Associates. Very high performance supercomputers at that time offered very good price performance; doing an excellent job of providing raw computing capability, they were highly competitive with less powerful machines in cost/computing capacity. As a result, they were used by commercial customers after having been pioneered for government users. But in those days all computers were, in fact, "custom" products: For each specific machine, the manufacturer typically designed both a special-purpose processor and a proprietary interconnect system linking that processor to the other components.

It was not until about the mid-1980s that Japanese firms entered the computer market, initially producing IBM-compatibles, then designs of their own, and ultimately machines that were quite competitive with those of Control Data and Cray. Around the same time came the very first wave of technological challenge from microprocessors, which made very small increments of computing power available to the end user in individual personal machines. Cray machines nonetheless remained quite cost-competitive into the 1980s.

Japan's Industrial Policy Creates Challenge to the United States

But what happened in that decade, a development that brought Dr. Flamm to the day's theme of innovation policy, was that Japanese technology advanced in a number of areas, including both the semiconductor and computer industries. It initiated the Fifth-Generation Computer Project and the Superspeed Computer Project, the latter being at least as important as the former even if less publicized.

For the very first time, Japanese producers were making significant inroads into the high-performance mainframe computer market.

This occasioned some alarm in the United States, particularly within the military and other government agencies that had originally funded computer technology. Given that superiority in information technology was seen as essential to the U.S. goal of having a qualitative technological edge in defense systems, it was of some concern that producers elsewhere were coming on stream with products that were competitive in demanding, high-performance applications.

One reaction was the Defense Advanced Research Projects Agency's (DARPA) launch in the 1980s of its Strategic Computing Initiative. The move was in part an attempt at responding to the technological challenges of the time, including competitive challenges from foreign companies producing systems that threatened to narrow the gap between U.S. information technology and IT available on the open market overseas.

Emergence of the Commodity-Based Supercomputer

Although DARPA program managers originally focused on custom components and on ways to use parallelism as an alternative design methodology to create new computer architectures, they gradually switched their emphasis to so-called commodity processors over the course of the initiative. The reasoning behind this change, which coincided with the arrival of the microprocessor on the industrial scene, was quite simple: Supercomputers were produced in relatively small quantities. Designing a custom processor to tweak the maximum possible performance, and producing it in relatively small scale, would result in a machine with an enormous price tag because all that R&D would be expensed over a relatively small number of units. But what if the new, microtype commodity processors that were being marketed in the millions, and whose costs were very much lower than those of custom processors, could somehow be harnessed? Even if they were less efficient at doing some of the calculations as individual processors, it might be possible to lash them together into a system and to figure out how to split up problems so that they could be handled by a large ensemble of cheap microprocessors networked together. That would prove a less costly way of getting the problem done, and, if scalable, could be used to solve any problem that was properly partitioned.

As a result, there emerged a whole new methodology for competing in supercomputers. Instead of focusing on the very high-performance individual processor, which was going to be enormously expensive to produce, a generic technology would be developed involving massively parallel systems that could run on relatively cheap components. Dr. Flamm characterized the strategy as a "kind of industrial jiu-jitsu": Rather than meeting directly the threat from "very, very well-done" high-performance processors coming out of Japan, the United States would "shift the terms of the battlefield." Many ultimately dismissed as a waste of money the $1 billion spent by the Strategic Computing Initiative between 1983 and 1993. Although numerous new firms appeared on the computing scene, many of them—even some that became major players, like Thinking Machines—ended up going out of business.

The other response to the Japanese challenge of the 1980s took the form of trade-policy initiatives, one of those being an attempt to open up Japan's market through forcing procurement by its government of U.S. supercomputers. In addition, dumping cases were filed in the United States in the mid- to late-1990s. Dr. Flamm pointed out that, between 1986 and 1992, the three major Japanese producers of noncommodity or "vector computers"—NEC, Hitachi, and Fujitsu—whittled the U.S. share of the market from nearly 80 percent to under 60 percent. "This threat was a very real threat," he remarked.

Federal Investment Transforms the Industry

Having set out the background, Dr. Flamm next discussed how investment by the United States in experimenting with a new set of technologies has ended up altering the competitive dynamics—the "industrial ecology"—of the supercomputer business. For it has indeed done so, and in some very significant ways that he did not believe to be widely appreciated, particularly in Washington. As a measure of industry dynamics he would use what was referred to as "Top 500" data, which described the 500 fastest computers in the world based on the LINPACK performance benchmark.[15] Not all computers were tested this way, and, he acknowledged, legitimate questions exist about whether the Linpack really is the best possible measurement of computer performance. Still, the numbers he would use were reflective of trends in and broader measures of the industry, and they were at once very easy to use and very detailed.

Arriving at the present, Dr. Flamm noted that it was with the inauguration of Japan's Earth Simulator, referenced in Dr. Kahaner's presentation, that the United States fell out of the world leadership in computing that it had held since about 1950. The United States had assumed the lead at that time from the United Kingdom, which had produced the very first electronic digital computers during World War II and continued in the No. 1 position until it "basically blew it" in the late 1940s. And the Earth Simulator had come on line not only as fastest in the world, but as faster by over a factor of two or three than its closest U.S. competitor. In 2002 many people compared the challenge this event presented to the United States, both technologically and in other respects, to the launch of Sputnik.

---

[15] According to Wikipedia, The LINPACK Benchmark measures how fast a computer solves dense n by n systems of linear equations $Ax=b$, a common task in engineering. The solution is based on Gaussian elimination with partial pivoting, with $2/3 \cdot n3 + n2$ floating point operations. The result is millions of floating point operations per second (Mflop/s). Accessed at *<http://en.wikipedia.org/wiki/LINPACK>*.

Growth of Industry Purchases Marks a Shift

Dr. Flamm then described a "major shift in the way the market works" that had taken place between June 1995 and December 2004. Historically, around two-thirds of supercomputers sold in the United States went to government, research users, or academic institutions. But, in slightly less than a decade, industry's share rose to the point that it was buying more than half of the most powerful machines on the market.

Meanwhile, there were no signs of a letup in the pace of computer performance improvement. Improvement was more or less continuous, with leading-edge outputs. In fact, the Earth Simulator, after reigning as the world's fastest machine for two-and-a-half years, has been displaced by an American machine, the Blue Gene L.

Dr. Flamm refuted the contention that there is a widening gap between the low and high ends of the Top 500, with the government continuing to acquire the truly fast machines and industry purchasing "shlock." The dispersion between the two ends, he said, is relatively constant with the exception of a big jump caused by the Earth Simulator, "an exceptional machine." Evidence does exist of a diminished industrial presence at the very top of the list, however: There has not been an industry-owned machine among the fastest 20 since about 2001.

United States Ascends, Japan Declines

Despite other trends over this period, the United States has enjoyed a "huge success story" in supercomputing. U.S. industry has emerged very strong, its share of Top 500 machines sold marching steadily upward while Japan's share is shrinking (Figure 27). Pointing to the top right-hand corner of his graph, Dr. Flamm specified that the small percentage of non-Japanese, non-U.S. machines indicated there represent products not from Europe but from China and India, the "new industrial powerhouses."

He added that the picture differs little if looked at from the point of view of total computing capability, which is in fact a better proxy for value. Furthermore, U.S. market share has been increasing not only worldwide but in each individual region of the globe, whether measured by number of machines sold or by total computing capability. Finally, U.S. manufacturers' share of the 20 fastest machines, after eroding during the mid-1990s, has emerged from the 1993-2004 period only slightly below where it had been at the start. The percentage of the Top 20 machines installed in the United States has followed a similar trajectory.

National Trade, Investment Policies Make an Impact

Evidence exists that national trade and industrial policies have had an effect on market behavior. In 2005 U.S. manufacturers of Top 500 machines controlled 100 percent of their home market, into which only a handful of Japanese

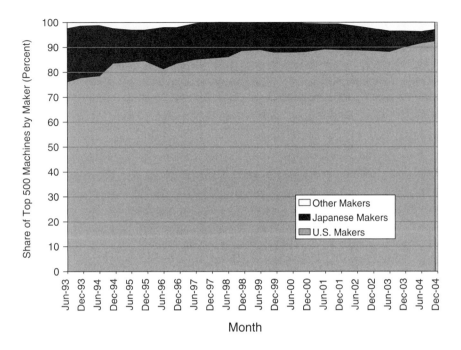

FIGURE 27 U.S. makers stronger than ever: Share of Top 500 machines (numbers) by maker.

machines in the category had been sold since 1998, and none at all since 2000. That contrasts with the situation prior to 1998, indicating that the dumping cases brought against Japanese manufacturers had made an impact. And while the U.S. share of Japan's market had dipped suddenly after the filing of the dumping cases, it popped back up again once they were settled. "It's hard not to think there's some causal connection there," Dr. Flamm observed.

Dr. Flamm next took up the thesis that the government-industry partnership formed to develop alternative methodologies for designing and building supercomputers has been quite a success and has transformed the nature of the supercomputer market over the previous decade. He noted that this is also the thesis of the National Academies report on which he has collaborated with a number of others, including computer scientists familiar with the industry.[16] He divided the architectures used in making supercomputers into three categories: "custom," applied to traditional machines with full-custom processors and full-custom interconnects between those processors; "commodity," denoting those

---

[16]National Research Council, *Getting Up to Speed: The Future of Superconducting,* op. cit.

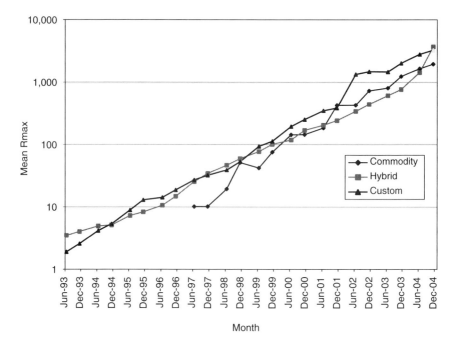

FIGURE 28 System performance by type: Mean R_{max} by system type.

made of microprocessors and interconnects available for purchase on the open market from third parties; and "hybrid," used for machines that generally had commodity processors but custom interconnects. "The very last segment of the computer industry to be transformed by the microprocessor and the PC has indeed been transformed," he declared, calling this "an extraordinary story."

Commodity Supercomputers Taking Over

The system performance of the custom and hybrid machines has improved almost in parallel since 1993, with that of commodity architecture tracing a similar slope since its advent in 1997, and there are no sign of any category's slowing down (Figure 28). Commodity machines' representation among the Top 500 has skyrocketed; however, from a standing start in December 1998, the commodity sector achieved an 80-percent share of all supercomputers sold by the end of 2004. In this rise, the commodity machines have essentially displaced the hybrids, which between 1993 and 1998 were displacing the full-custom machines. The change is nearly identical when tracked on the basis of capacity

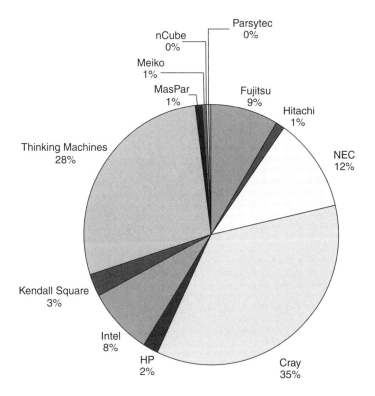

FIGURE 29 Restructuring of product has restructured industry: Top 500 market share (R_{max}) by company, June 1993.

rather than units, and differs only slightly when consideration was narrowed to the Top 20 machines.

Transforming a Market Restructures the Industry

This transformation of the supercomputer market has completely restructured the computer industry as a whole. According to a graph illustrating the composition of the market for the Top 500 by computing capacity (Figure 29), the two leaders as of June 1993 were Cray, with 35 percent of the market, and Thinking Machines, another U.S. company, with 28 percent. Japan's NEC, Hitachi, and Fujitsu combined for a 22-percent share, as compared to the more than 40 percent of the vector-supercomputer market by machines installed that they enjoyed in 1992.

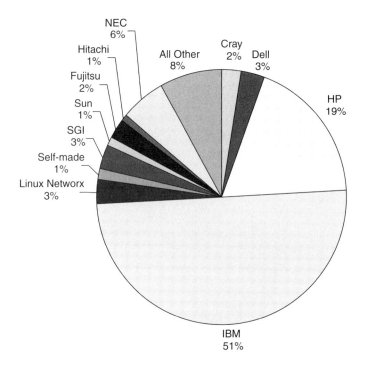

FIGURE 30 A whole new playing field: Top 500 market share (R_{max}) by company, June 2004).

By June 2004, the Top 500 market was radically altered (Figure 30). IBM, out of the previous picture entirely, was the leader with more than half of all capacity sold; H-P had climbed into the No. 2 spot, its 19-percent share almost an order or magnitude higher than the 2 percent it had held 11 years earlier. While NEC was the third-largest player at 6 percent, the total share of the Japanese manufacturers was only 9 percent. Cray's share of the market had plunged to 2 percent.

Government-Only Demand for Traditional Technology

Traditional custom supercomputers, still required by government users for some applications, are showing signs of becoming "a government-only island," as illustrated by a graph which indicates that around 60 percent of custom machines consistently go to purchasers in the government and research sectors (Figure 31). From the perspective of computing capacity, the figures are even more one-sided: Since 2001, 85 percent of the market for custom systems has

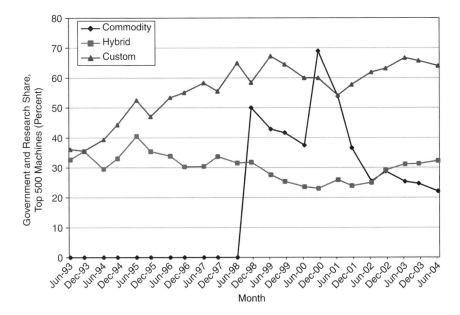

FIGURE 31 Custom supercomputer market becoming a government-only niche: Government and research share, top 500 machines.
NOTE: Balance is industrial, vendor, academic.

been accounted for by government and research users. "A strategy started by the military has become so dramatically successful that some government users—a minority, fortunately—are essentially cut off from this whole new development in computing technology," observed Dr. Flamm, likening the current market for custom supercomputers to those for submarines and aircraft carriers.

He then presented a number of conclusions:

- The conventional wisdom holding that the government role in computers is much diminished has never been true at the high end and is still not true. Critical government applications have motivated policy in that area.
- The policy implemented in the 1990s in response to the challenge that U.S. government users faced in the previous decade have proved to be a huge success, even though the ultimate game plan has not matched the original. The technological foundations residing in "those dead companies that littered the landscapes" have fueled new ideas and new methods that have led to industrial outcomes that are highly favorable to the United States.

- With considerable evidence available that the custom-supercomputer market is increasingly a government-only niche, the government might be forced to view it the way it views weapons systems and to pay for the full load if its need for the product persists.
- Spillovers from supercomputers seem to have slowed, as has super-computer R&D, and "spin-on" has become as important as spin-off. This, again, is one of the messages of the National Academies report to which Dr. Flamm has contributed.
- It is hard to see how trade barriers have helped either U.S. supercomputer producers or users.
- The government's challenge is to maintain whatever capabilities it needs that are outside the commercial mainstream while leveraging developments in the industry it has so successfully transformed in such a way as to make cost-effective solutions linked to the mainstream accessible to government users.

In closing, Dr. Flamm held up the history he had recounted as "an example of a government-industry partnership in technology development that has yielded unforeseen but impressive results as an industrial outcome for the United States."

Dr. Edelheit, thanking Dr. Flamm and moving on to the next presentation, stated that best practice in government-industry-university collaboration was to be found not abroad but at the U.S. National Institute of Standards and Technology in the form of the Advanced Technology Program (ATP), whose director, Marc Stanley, would follow.

CROSSING THE VALLEY OF DEATH:
THE ROLE AND IMPACT OF THE
ADVANCED TECHNOLOGY PROGRAM

Marc G. Stanley
National Institute of Standards and Technology

Mr. Stanley, thanking Dr. Edelheit and expressing his pleasure at attending the symposium, remarked that, speaking later in the day, he benefited from having heard the earlier discussion of how countries might profit from public-private partnerships. The observations he would offer, he said, would not necessarily be limited to the ATP.

Conceived by two congressional staff members, one in each house, the ATP mission is "to accelerate the development of innovative technologies for broad national benefit through partnerships with the private sector." Mr. Stanley described societal benefit as "a very interesting criterion" for a public-private partnership to use when considering investments in high-risk technology. The key to the process was getting industry to lead it, he said. His answer to one of the questions most frequently asked about ATP—was government actually capable of making this kind of investment?—was that more than a dozen years of observing the program had taught him that industry knows best where particular market gaps might be.

U.S. Market's Total Autonomy a "Myth"

While acknowledging that the U.S. market is very open and lacks the regulatory constraints often seen overseas—and that these exigencies both provide strength and act as a motivator—Mr. Stanley likened to a "myth" the belief held both at home and abroad that, in the United States, markets acted entirely on their own. Impediments to early-stage, high-risk investment remain a cause for concern in the United States; at the same time, the government's role in technology development is underappreciated. He would therefore devote some of his time to the question of whether there is a role for government to play in innovation.

To set the stage for this discussion, Mr. Stanley referred to a plaint against Roman rule from *Monty Python's Life of Brian* that, for *The Economist* (May 1, 2004), had evoked criticisms heard from Americans of their government's role in ushering major technological innovations: "But what, apart from the roads, the sewers, the medicine, the Forum, the theater, education, public order, irrigation, the fresh-water system and public baths . . . what have the Romans done for us? (And the wine, don't forget the wine . . .)." Placing this debate in a U.S. context, he recalled that Alexander Hamilton had created the nation's first public-private partnership, a planned industrial center in New Jersey called the Society for Establishing Useful Manufactures (SEUM), as a tool for competing with Great Britain.

Partnerships a Hallowed U.S. Tradition

The history of U.S. partnerships provides ample evidence of government involvement apart from that just supplied by Dr. Flamm in his discussion of the computer industry. To cite a few examples:

- **1798**—Congress provides a grant for production of muskets with interchangeable parts to Eli Whitney, who founds first machine-tool industry
- **1842**—Samuel Morse receives award to demonstrate feasibility of telegraph
- **1919**—RCA founded on initiative of U.S. Navy with commercial and military rationale (patent pooling, antitrust waver, equity contributions)
- **1969-1990s**—Government investments develop the forerunners of the Internet
- **present**—The government is currently making major investments in genomic/biomedical research

Mr. Stanley asked the audience to focus on two facts: 1) that deep capital markets exist in the United States, but 2) that some underinvestment in pre-competitive technologies nonetheless remains. He would attempt to reconcile these facts and, in the process, demonstrate that, under the proper circumstances, public-private partnerships can play a key role in helping countries anywhere in the world to compete.

Private Investors Neglect Early Stage

To begin, he raised a question: If the United States has large and well-developed early-stage capital markets—indeed, the world's best—thanks to broad angel markets and deep venture markets, what is the issue? In answer, he noted that of about $20 billion in rounds of venture-capital investment in 2004, only $105 million represent seed rounds, a constant trend in recent years . Another graph (Figure 32) showed that the median deal size for early-stage seed-round investments had fallen to $300,000 by the end of 2004 from $1 million at the beginning of 2001. Much more of the private equity community's money, therefore, is going into later-round investments, which typically fund such later-stage business activities as product development and marketing. In addition, the distribution of venture capital tends to concentrate in a limited number of geographical regions.

Mr. Stanley listed several reasons for what he contended was underinvestment in precompetitive technologies in the United States:

- Markets, although powerful, are imperfect.
- New ideas lack constituencies.
- Venture capitalists tend to invest later in the cycle.
- Firms can't capture the entire value of some investments when acting alone.

ATP's Role More Extensive Than Recognized

A review of the funding of early-stage, high-risk research conducted by Lewis Branscomb and Philip Auerswald of Harvard's Kennedy School under contract to ATP has produced estimates placing investments by venture-capital firms, state governments, and universities at only 8 to 16 percent of the total funding applied to early-stage technology development. The federal government, through ATP and the Small Business Innovation Research (SBIR) program, is estimated to account for between 21 and 25 percent of these moneys (Figure 33). The study's conclusion is that ATP plays a greater role in financing the development of precompetitive technology than is widely appreciated.

Outside the United States, as suggested both by some of the day's earlier speakers and by Mr. Stanley's own travels in member countries of the Organisa-

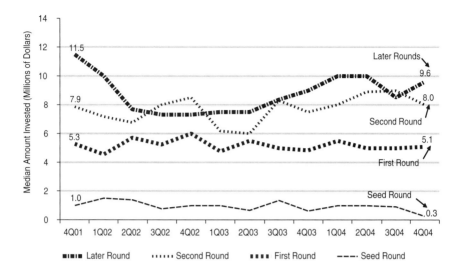

FIGURE 32 Early-stage deal size declines: Median amount invested by round class.
SOURCE: Adapted from Dow Jones Venture One/Ernst & Young.

tion for Economic Co-operation and Development (OECD), there has been a "clarion call" for enlisting government involvement as one tool for building a dominant position in certain areas of technology. To illustrate, Mr. Stanley displayed a comment by Elizabeth Downing, an official of the ATP award winner 3D Technology Laboratories: "Why should the government fund the development of enabling technologies? Because enabling technologies have the potential to bring enormous benefits to society as a whole, yet private investors will not adequately support the development of these technologies because profits are too uncertain or too distant."

In a similar vein, the noted venture capitalist David Morgenthaler has remarked that "It does seem that early-stage help by the government in developing platform technologies and financing scientific discoveries is directed exactly at the areas where institutional venture capitalists cannot and will not go." And Jeffrey Schloss, when speaking on behalf of Dr. Francis Collins of the National Human Genome Research Institute of the National Institutes of Health, has said: "The Advanced Technology Program can stimulate certain sectors like biotechnology where the risk is such that the private-sector investment is ineffective or nonexistent. Because of its synergies across a broad range of technologies, ATP has advanced the research being done in DNA diagnostics tools."

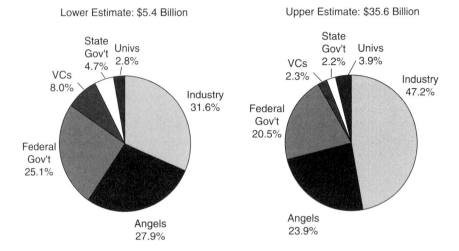

FIGURE 33 Early-stage technology development: Estimated distribution of funding sources for early-stage technology development, based on restrictive and inclusive criteria.

NOTE: The proportional distribution across the main funding sources for early-stage technology development is similar regardless of the use of restrictive or inclusive definitional criteria.

SOURCE: Lewis M. Branscomb and Philip E. Auerswald, *Between Invention and Innovation: An Analysis of Funding for Early-Stage Technology Development*, NIST GCR 02–841, Gaithersburg, MD: National Institute of Standards and Technology, November 2002, p. 23.

Market Inefficiencies Make Federal Role "Critical"

On the basis of such statements, as well as other evidence, one might infer that the federal role in dealing with the investment gap affecting early-stage, high-risk technologies is critical, Mr. Stanley remarked. Markets for allocating risk capital to early-stage technology are not efficient, he asserted, due to inadequate information for making investment decisions, the high uncertainty of outcomes, and difficulty in appropriating the benefits of early-stage enabling technologies. As Dr. Branscomb has said in testimony before the U.S. Senate Committee on Commerce, Science, and Transportation, "Entrepreneurs and private equity investors alike consistently state that there exists a financial 'gap' facing early-stage technology ventures seeking funding in amounts ranging roughly from $200,000 to $2 million."

Foreign competitors, meanwhile, have technology-support programs larger than those of the United States, and those programs employ a broad range of measures in such domains as trade, tax, procurement, standards, government equity financing, and regional aids. To illustrate, Mr. Stanley offered a few details from programs discussed earlier in the symposium. Finland's Tekes runs a program that was similar to ATP but is financed at an annual level of around 387 million euros as compared to ATP's $140 million. Japan's ASET program, one of six Japanese semiconductor partnerships under way, received $473 million in the period 1995-2000. In the European Union, JESSI was funded at $3.6 billion for the period 1988-1996; MEDEA+ was receiving 500 million euros annually; and 17.5 billion euros is to be provided under the Framework Program over 5 years. In addition, not only Canada but other nations of the OECD are practicing many forms of public-private partnerships to compete with the United States.

Foreign Competitors' Concerted Efforts in R&D

Posting a chart that traces over time a number of developed nations' R&D expenditures in percentage of GDP, Mr. Stanley remarked that it should not be surprising that the trend has been upward. The data graphed is based on a wide variety of programs, including direct grants, loans, equity investments, and tax deferral; whether the investments represented include both civilian and dual-use technology, or only the former, has not been determined. "Clearly, other countries are increasing their expenditures on R&D and have taken up concerted efforts," he stated.

Mr. Stanley then turned to ATP's workings. Unlike SBIR, some of whose funding was agency- or mission-specific, ATP has a collaborative focus and a flexibility under which funding is available to all technologies, enabling industry to address large problems. Cost sharing, which combines private and public funding, and serious commitment to commercialization are requirements. He supplied a list of attributes that he considers "pillars to develop a good public-private partnership," saying each contributes to success, both within ATP and abroad: These are

- an emphasis on innovation for broad national economic benefit;
- a strong industry leadership in the planning and implementation of projects;
- project selection based on technical and economic merit;
- a demonstrated need for funding;
- a requirement that projects have well-defined goals;
- provisions limiting the funding period;
- a rigorous competition based on peer review;
- encouragement of collaboration among small, medium-sized, and large companies; universities; and clusters and industry parks; and
- evaluations, to be pursued both by presenting results to the granting agency and by establishing a baseline as an aid to proving success.

Measuring ATP's Success in Societal Benefits

How successful has ATP been? The program was able to show net societal benefits of $17 billion based on the analysis of only 41 of the 736 projects it has funded; the total cost to the federal government over the life of the program was about $2.2 billion. "This is just the beginning," Mr. Stanley promised. "The rewards are continuing to come in."

But success is not to be measured exclusively in return on investment. The White House Office of Management and Budget, following a review according to its Performance Assessment Rating Tool (PART) 2 years before, praised ATP's assessment effort, and a National Academies review had called it "one of the most rigorous and intensive efforts of any U.S. technology program."[17]

Evaluation Considered an Integral Component

ATP's selection process, monitoring, and follow-up on projects is "exceptional," Mr. Stanley asserted, adding that the program has the ability to identify unsuccessful projects and that those projects are terminated. "You have to terminate companies that are not successfully doing what they say," he commented. "And then you should be able to speak not only of your successes but of your failures, because there are lessons to be learned from that."

ATP used a number of methods to measure the activity of its awardees against the program's mission:

- **Inputs:** ATP funding; industry cost-share;
- **Outputs:** R&D partnering; risky, innovative technologies; S&T knowledge;
- **Outcomes:** acceleration; commercial activity; and
- **Impacts:** broad national economic benefits.

The program had in the previous year published a book on its measurement tools, and the book had "become a very hot seller overseas."

To conclude, Mr. Stanley provided examples of the kinds of studies ATP has employed in order, as he put it, "to do the kind of work that we think is essential to maintain fidelity to the taxpayer[s] for the investments that they've given us":

- **statistical profiling** of applicants, projects, participants, technologies;

---

[17]According to the Office of Management and Budget, "the PART was developed to assess and improve program performance so that the federal government can achieve better results. A PART review helps identify a program's strengths and weaknesses to inform funding and management decisions aimed at making the program more effective. The PART therefore looks at all factors that affect and reflect program performance." Office of Management and Budget Web site, accessed at <*http://www.whitehouse.gov/omb/part/*>.

- **status reports** (mini case studies) for all completed projects;
- **econometric and statistical studies** of innovation and portfolio impacts;
- **special-issue studies**;
- **progress tracking** of all projects and participants (business reporting system, other surveys);
- **detailed microeconomic case studies** of selected projects, programs;
- **macroeconomic impact projects** from selected microeconomic case studies; and
- **development and testing** of new assessment models, tools.

The program reports on these extensive studies, he said.

Dr. Edelheit thanked Mr. Stanley and introduced Pace VanDevender to talk about how the Department of Energy's National Laboratories fit into this landscape.

SANDIA NATIONAL LABORATORIES: DOE LABS AND INDUSTRY OUTLOOK

J. Pace VanDevender
Sandia National Laboratories

Expressing his pleasure at participating in the symposium, Dr. VanDevender specified that he would present the legislative basis for the national laboratories' involvement in technology transfer in light of the fact that their mission was a governmental function (making the United States safe and secure). He said that he would also explain how technology partnerships support DoE in that function, discuss the competitive advantage to industry of collaborative research and licensing and, finally, proffer some closing remarks.

Dr. VanDevender explained that DoE National Laboratories are government-owned, contractor-operated organizations (GOCO). Under this arrangement, the labs' property and equipment belongs to the government, and their people and reputation are affiliated with a contractor. The labs' missions, however, are fixed by Congress and range from the pursuit of knowledge to the maintenance of nuclear weapons. In fiscal year 2003, the laboratories received around $6 billion of DoE's $8.5 billion R&D budget. "It is a lot of money," he remarked, "and every dollar has its mission."

DoE's Tech-Transfer History a Quarter-Century Old

Dr. VanDevender then tracked the policy basis for DoE's involvement in technology transfer. This began in 1980, when the *Stevenson-Wydler Technology Innovations Ac*t, which established technology transfer as a mission for the federal laboratories. Each lab set up an Office of Research and Technology

Applications to help disseminate information: "If it wasn't classified, we were to publish it," he recalled, "and if it was useful, industry could use it." This legislation also established a preference for U.S. manufacturers, something that persists to the present, "even though," as he noted, "globalization has radically changed the nature of the industrial world."

A second major landmark came in 1984 with the *Trademark Clarification Act*, which gave the GOCOs licensing and royalty authority for the first time. Then, in 1986, the *Federal Technology Transfer Act* extended the responsibility for technology transfer to lab employees, so that each individual's performance evaluation took into consideration fulfillment of this mission. "That did not work at all, as you might imagine," Dr. VanDevender recalled. "It just didn't engage the consciousness of the lab employees."

A turning point was reached with the *National Competitiveness Technology Transfer Act of 1989*. This act made technology transfer a mission of the DoE weapons labs. It also allowed GOCOs to enter into cooperative research and development agreements (CRADAs): *i.e.,* to make deals with industry that involves becoming partners and cofunding R&D activities. Also in 1989, the *NIST Authorization Act* recognized CRADA intellectual property other than inventions and, thereby, helped resolve a problem that had inhibited technology transfer from the labs. The *1995 National Technology Transfer Act* guaranteed to industry the ability to negotiate for rights to CRADA inventions and increased the royalty distribution limit that had been placed on lab inventors, thereby increasing their motivation to invent.

Assessing the Efficacy of Tech-Transfer Policies

How well has this policy basis for GOCOs' participation in technology transfer worked? "Pretty well," Dr. VanDevender assessed, posting a tabular account of the DoE labs' tech-transfer activities for fiscal year 2004 that displayed what he called "some respectable numbers" (Figure 34). However, the approximately 10,000 tech-transfer actions that took place during that year across DoE's entire complex of 24 National Labs and other facilities were "not at all uniformly distributed." In fact, about half of the activity, as measured by funds coming in from industry, was attributable to a single laboratory out of the 24. On a lab-to-lab basis, therefore, tech-transfer activity had been "fairly modest."

Selecting among the figures, Dr. VanDevender noted that there were 610 active CRADAs across DoE in FY2004; based on the fact that 157 had been initiated in that year, he posited an average lifetime for CRADAs of 3 to 4 years. Whether the 520 patents issued in FY2004 amounted to "a lot or a little," he allowed, "depends on your perspective." Of 4,345 licenses active in FY2004, 616 were new and 3,236 were income bearing. Agreements classified under "active work for others, nonfederal entities" (WFO/NFE), numbered 1,884, while those classified under "active work for others, other federal agencies" (WFO/Other

• More than 10,000 technology transfer actions in FY 2004	• CRADAs — Active CRADAs — New CRADAs	FY 2004 610 157
	• Intellectual Property	
• Incorporates activities across DOE complex of 24 national labs/facilities	— Invention Disclosures — Patent Applications — Issued Patents	1,617 661 520
	• Licenses	
• Supports DOE/NNSA missions by enhancing lab capabilities and commercializing technologies	— All Active Licenses • New Licenses — Patent Licenses — Active — Other IP Licenses — Active — Income-Bearing Licenses	4,345 616 1,362 2,983 3,236
	• Active WFO/NFE Agreements • Active WFO/Other Fed. Agreements • User Facilities Agreements/ FY04	1,884 2,782 3,252

FIGURE 34 Technology transfer supplements the primary mission of each lab.

Fed. Agreements) numbered 2,782. User-facilities agreements came to 3,252 in FY2004.

Results Plateauing in Numerous Areas

Dr. VanDevender then projected a graph showing that several measures of activity in the intellectual-property domain has plateaued—i.e., very little change with time—as technology-transfer policy has evolved (Figure 35). Invention disclosures, after growing vigorously in the decade ending in 1997, dipped sharply and then, around the close of the century, hit something of a plateau. Patent applications and patents awarded, moving almost in lockstep with one another, have followed a similar if somewhat steadier course.

CRADA activity shows a rapid increase from 1992 to 1996, after which it drops off significantly (Figure 36). Dr. VanDevender's explanation was that when CRADAs came into being, they were cofunded by government and industry, with a 50-50 share being typical. When the government participated as a "funds-producing partner," CRADAs grew rapidly. "But when the economy recovered and there were other pressing needs for federal money," he said, "the federal matching dollars went away, and much of the CRADA activity did also." It has not, however, fallen to zero, because some robust partnerships were established during the years of upward growth and these industry participants provided the funding needed to sustain their CRADAs.

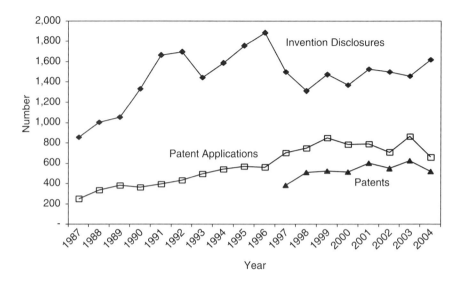

FIGURE 35 Invention disclosures and patents have plateaued under current policies and priorities.

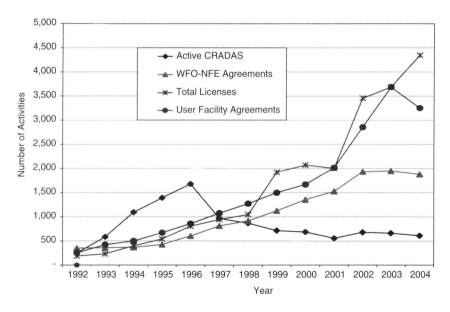

FIGURE 36 Industry has preferred other vehicles over CRADAs without matching funds.

WFO/NFE agreements also reached a plateau in 2002 after many years of sustained growth. However, licenses set a pattern of swift growth that is continuing, and agreements covering user facilities appear to have done likewise until their sharp decline in 2004. "The message from both of those," Dr. VanDevender reflected, "is that there's been a lot of growth, a lot of deals were made and relationships built, but it has really plateaued under the current policies."

Gauging Tech Transfer's Benefit to the Taxpayer

The benefit of this technology-transfer activity to the U.S. taxpayer is substantial, according to Dr. VanDevender. A joint program with Goodyear Tire and Rubber Company in computational mechanics and predictive reliability is an example of a fruitful endeavor for Goodyear and for national security. How could research involving tires be of value to the nuclear weapons program? "If you think of the nose cone of a B-61 crashing into the earth, it is a highly deformable material problem," he explained. "Tires rotating on your car are continually deformed. Computer codes that were developed at Sandia for the B-61 were extended in partnership with Goodyear." The resulting benefit for the company is the Assurance tire, which surpassed its 2004 sales projection of one million units by 100 percent; the benefit for DoE is a better finite element code for government applications.

Figure 37 provides examples of CRADA projects that benefit a corporation and "often gets products into the hands of the public to protect us all" conveying "the gist of dual-use."

Meanwhile, DoE's income from licensing increased over time to $27 million in 2004 (Figure 38). Other DoE intellectual property income, from copyrights and other sources, have lagged far behind that. The extra money was "helpful," Dr. VanDevender said, in that it made DoE "more agile in meeting [its] needs." About 20 percent of the licensing income typically went to the inventors and the rest was put to a variety of uses, such as:

- upgrading the Advanced Photon Source at Argonne National Laboratory;
- establishing the Technology Maturation Fund used by many of the DoE labs;
- supporting startups through the Center for Entrepreneurial Growth;
- developing the fan airfoil for improved energy efficiency; and
- reinvesting in teams that had developed and licensed the intellectual property.

"Think of [the last use] as very early-stage seed money," he suggested, noting that it was usually distributed in amounts no larger than $300,000 and often as small as $10,000. Still, such a "microinvestment . . . gets teams started [and] makes them more credible so they can attract other teams."

- **National Security**
  - Computational mechanics and reliability with Goodyear Tire and Rubber Company
  - Explosives detection with U.S. Industry Coalition (USIC)
  - Synthetic aperture radar with General Atomics
- **Energy Security**
  - Technologies for light-duty trucks with U.S. Advanced Battery Consortium (USABC)
  - Lithium battery technologies with USABC and other industrial partners
  - Risk assessment of geologic carbon dioxide with BP AMOCO
  - Electric efficiency & reliability with California Energy Commission
- **Scientific Research**
  - Advances in structural biology with various industrial partners
- **Environmentally Sound Energy Technologies**
  - Advanced turbine systems with Siemens Westinghouse Power Corporation

FIGURE 37 "Funds-in" agreements advance DoE/NNSA objectives.
NOTE: Representative examples are cited.

Industry Benefiting Enough to Stay Engaged

Industry, for its part, is gaining sufficient competitive advantage through such collaborative research and licensing activities to stay engaged with DoE, Dr. VanDevender said. He cited a variety of outcomes:

- A major manufacturing company has reduced design time and eliminated the need for multiple prototypes with codeveloped simulation tools.
- Chemical and power companies have improved processes and plant designs using an advanced software toolkit.
- Equipment able to detect radiation from high-speed vehicles—potentially useful in identifying terrorists in possession of nuclear materials—has become available to the public sector through licensed suppliers.
- A new duct-sealing system that enabled home and business owners to save energy has been a deployed through more than 60 commercial franchises.
- A company has developed cancer treatments based on licensed intellectual property.
- Airfreight containers employing advanced materials developed through collaborative research has gone onto the market.
- A commercially successful family of hydrogen sensors has been licensed to help promote the future hydrogen energy distribution system.

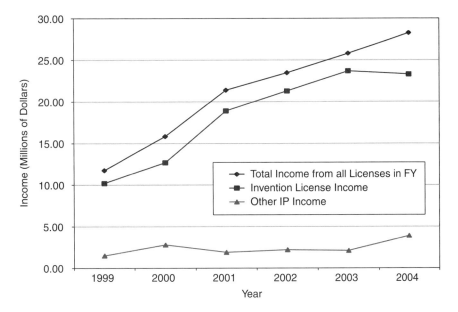

FIGURE 38 Licensing income supplements the $6 billion of federal funding at labs to enhance mission results.

Sandia's S&T Park Growing Robustly

In contrast to CRADAs, patent applications, and other tech-transfer vehicles whose growth has plateaued, science and technology parks are springing up as a new thrust. Sandia National Labs Science & Technology Park (SS&TP) in its seventh year, had drawn $167 million in investment—$146.6 private, $20.4 million public—and was still growing. A pedestrian-oriented, campus-style installation located on a tract of land exceeding 200 acres in area, SS&TP in the spring of 2005 housed 19 organizations with 1,098 employees that occupies almost 500,000 square feet. Sandia National Laboratories provides redundant power and state-of-the-art connectivity to the park, plus it helps tenants accelerate city approval processes. Tenants paid in $17 million to Sandia Labs while acquiring contracts from the labs worth $85.6 million; "the government, Sandia, and industry," observed Dr. VanDevender, "therefore benefit as funds flow both ways." Projections put SS&TP's final size at about 2.3 million square feet and its final workforce at between 6,000 and 12,000.

Dr. VanDevender recalled visiting Dr. Chu at ITRI a few months before the symposium and seeing Hsinchu Science Park, which produced about 10 percent of Taiwan's GDP. "We have 1,000 people, they have 100,000, but they're 30 years along," he said. "Obviously, there's much we can learn from them, and we're interested in doing that and have begun that relationship." Contrasting

ITRI's and DoE's models, he noted that the former is based on a single-purpose mission of technology development and commercialization with relationships, while the main mission of the DoE labs is "national security broadly writ"; for ITRI, therefore, technology transfer is a dedicated mission, whereas for DoE it is a supplementary mission.

Comparing Results of DoE, ITRI Models

The DoE labs receive about ten times as much annual funding as ITRI, or $6 billion versus $600 million. But industrial contributions account for only about $60 million of the DoE labs' funding, or 1 percent, while around $200 million, or one-third, of ITRI's funding comes from industry. The latter is "a very powerful statement of support consistent with their mission," Dr. VanDevender said.

The DoE labs produce around 600 patents per year, half as many as ITRI's 1,200; this translates to 0.1 patent per $1 million for the DoE model and two patents per $1 million for the ITRI model. However, the difference in patents per industry dollar is far narrower—about 10 patents per $1 million from industry for DoE versus six patents per $1 million from industry for ITRI—because industry is leveraging the huge U.S. investment in national security in the DoE model. However, these two rates are "very comparable," said Dr. VanDevender, "given the uncertainty in the value of those patents, [and] particularly since many more companies are spun off from ITRI than from DoE labs." Both models have their strengths and both were valuable, he concluded, suggesting that the comparison raised a question worth considering at the next stage of policy making: "whether or not [the United States] should experiment with a single-mission lab for industrial competitiveness."

In closing, Dr. VanDevender affirmed that technology partnerships add significantly to the innovation capabilities of the DoE and NNSA labs, as well as to the innovation capabilities of their industrial partners. They expand the R&D capacity for both industry and the labs and contribute to the fulfillment of DoE's mission and goals while providing competitive advantage to those industry partners that have long-term relationships with the laboratories. Although the effect is very small in percentage terms, the partnerships continue to provide important results for the government's national security mission (even during very constrained federal budgets) and differentiating benefits for the partnering corporation. Nevertheless, he emphasized that partnership activities have "really plateaued under the current policies and priorities; a wiser and bolder approach is needed to move partnerships to the next level of effectiveness and efficiency."

Discussion Roundtable:
What Are the Conditions for Success?

Moderator:
Mark B. Myers, The Wharton School

Stefan Kuhlmann, Fraunhofer ISI, Germany
Hsin-Sen Chu, Industrial Technology Research Institute (ITRI), Taiwan
Peter J. Nicholson, Office of the Prime Minister, Canada
Marc G. Stanley, National Institute of Standards and Technology
Lewis S. Edelheit, General Electric, retired

Dr. Myers explained that rather than asking the members of the final panel to speak, he would give the audience an opportunity to pose the questions that, he was sure, were accumulating as the day progressed. First, however, he turned to a fellow panelist, Dr. Edelheit, who indicated at the close of the previous session that he had numerous questions in mind.

MIGHT ATP SPONSOR PROGRAMS IN
FOCUSED RESEARCH AREAS?

Dr. Edelheit, addressing Mr. Stanley, remarked that the presenters from other nations had all singled out areas that they viewed as important to their respective countries or companies in those countries and in which work is therefore focused. ATP, although the closest analogue to the foreign programs under which such work was taking place, did not seem to sponsor work in focused areas. He asked what Mr. Stanley thought about that issue and whether, in fact, he views it as an issue that should be considered.

Mr. Stanley responded that there were several ways to address the issue. The National Academies, in a review of ATP published 2 years or so before, recommended that ATP look at the United States' highest priorities and create opportunities for thematic competitions in sectors that are either important to national security or technologically relevant for the nation's future. The program's funding has, however, been held at a reduced level for a number of years, and it remains all but impossible to run focused competitions of this sort in the absence of "a certain baseline of allocation."

Nonetheless, ATP management has, in observing submittals to its regular competitions, seen proposals that concentrate on particular areas that industry feels to be both important and relevant for the research of the ensuing 8 to 10 years, and ATP has succeeded in fashioning "virtual focused competitions" in those areas. For example, ATP has received a suite of proposals indicating a great deal of interest in optoelectronics. Similarly, new work has been identified in the areas of homeland security and technologies for assisted living and care for the elderly. Around 7 years earlier, ATP had started receiving proposals dealing with how the United States could improve its position in fuel cells, hydrogen, and various other alternative fuels. Thus, industry is using ATP in a variety of ways, and the program's management can collect this data and show where the program fares in terms of focused competitions.

COULD INCREASED FUNDING CHANGE
ATP'S SELECTION METHODS?

Dr. Edelheit then asked whether Mr. Stanley believed that, in the event that its funding rose, ATP would try to use at least some of the increase to develop areas of focus that made sense.

Mr. Stanley said that this question was a difficult one for him to answer: He worked for the President of the United States and so had to abide by his budgetary proposal to Congress, which calls for the elimination of ATP in fiscal year 2006 in favor of other national priorities. "In the event that were to change by some congressional disposition of funds, and the amounts were appropriate," he acknowledged, "then certainly we could consider that. But, at this particular point, I have to wait and see what happens."

CANADA'S MARKET: HOW MUCH POLICY DIRECTION?

Dr. Myers then directed a question to Dr. Nicholson. He recalled that Dr. Marburger had talked about the strength to be drawn from the diversity of the U.S. system, and he commented that that statement reflects what had been the view of the U.S. government for some time: "that we do not have an industrial policy, we do not pick areas of winners [and] losers." But many of the day's presentations had indicated that China, Japan, and Europe are clearly choosing which industries to support and have thereby drawn a distinction between two kinds of policies. As Canada, in his observation, tended always to fall "somewhere in between," he wished to know the Canadian thinking on the subject.

Dr. Nicholson, calling Dr. Myers's question "terrific," declared that Canada was indisputably "in the grip of schizophrenia." The philosophy in the country, at its current state of development, is "very much to be guided by the market—in other words, a market pull rather than a technology push." That said, however, Canada does not have as much native diversity as the United States simply because of the scale of its markets, and circumstances do matter in thinking about strategy.

If a sector orientation does exist, and Dr. Nicholson sensed that one was developing, it is probably in the area of energy and environmentally related technologies. The former is obviously of great concern to the world, and Canada enjoys a relative abundance of energy resources. As for the latter, any projection of the world's future needs will include increasing demand for ways to lower the carbon content, or the pollution content, of units of GDP.

ENVIRONMENTAL TECHNOLOGY A SPUR TO INNOVATION

Regarding environment as an emerging priority, Canada has created a couple of foundations devoted to it and has recently released a plan for meeting the Kyoto targets for reducing greenhouse gases sometime in the 2010-2012 period. Although admitting this is "a very stretchy target," Dr. Nicholson called it "probably the toughest one in the world relative to the business-as-usual trajectory that we've got to cope with." He added that whether or not the target is met, setting the target will generate new demands as some of the large emitters try to meet their specific targets. In turn, this will stimulate innovation in Canada, he predicted, and as "all kinds of smart people and entrepreneurs" compete for a significant amount of money available through the new Climate Fund.[18]

This is one of numerous instances in which the benefit of a policy that was established for rather different reasons might end up having the happy, unintended consequence of being a big driver for innovation. It might also prove "a bit of an equalizer vis-à-vis the [United States]," Dr. Nicholson said, suggesting that the incentive is about to burgeon in Canada "to really work hard on producing green energy that can be exported in its technological form around the world."

---

[18]Through a new Climate Fund, the Government of Canada intends to purchase 75–115 Mt of reduction credits a year, up to 40 percent of the total reduction needed in 2008–2012. The government agreed to allocate CAD$1 billion per year over the next 5 years and projects funding of $4 billion–$5 billion 2008–2012. Pew Center on Global Climate Change Web site, accessed at <*http://www. pewclimate.org/policy_center/international_policy/canada_climate_plan.cfm*>.

BASIC RESEARCH: HOW IMPORTANT TO INNOVATION?

A questioner from the audience, who identified herself as a Swedish science reporter temporarily working at *Science Magazine* in Washington, observed that most countries are experiencing a decrease in governmental funding for basic research and asked for assessments of what that could mean for the future of innovation. Is basic research important for innovation, and does it need to be funded by governments, or are there other possible sources?

Dr. Edelheit, while first acknowledging that complexity surrounds the issue of "basic research versus applied research," stated that the answer depends on the country and on the economic drivers for R&D operating there. The United States remains the world leader in basic research by far, with other countries—and he pointed specifically to developing countries—doing much less "real basic research." But in the end, he predicted, there will be a balance. As other countries become strong enough that they can afford it, they would in their own best interests start doing more basic research. What will happen in the United States, in contrast, is that funding for R&D will decline and some basic research will be converted to solving national needs more directly in the interest of speed.

Recalling that Dr. VanDevender had talked about Sandia's changing its model, and adding that industry has certainly changed its model, Dr. Edelheit remarked that the National Institutes of Health and ATP were beginning to go in the other direction. This shift might go too far and eventuate in a balance, he conjectured, adding: "There always needs to be a balance." He also expressed his belief that too much basic research is being done in the United States relative to the kind of work on which the day's discussion had focused.

DO U.S. FIRMS NEGLECT THE PUBLIC GOOD?

Dr. Myers said that he, like Dr. Edelheit, was a former corporate head of research, in his own case at Xerox, but that he had a somewhat different concern. For a period of time the United States has been an important source of information that, having come out of the private sector through "quasi-national research laboratories" such as Bell Labs, IBM Yorktown, GE, and Xerox PARC, "went into the public good." One characteristic of that time was that all of the companies he had named were operating as monopolies. "Every industry loves a monopoly," he remarked, saying that a "sort of openness" was possible in that era.

While the corporate laboratories have not disappeared, the public-good functions they performed is no longer present in any of their organizations. This has not happened solely because the companies have lost their monopolies, he contended, observing that Microsoft, Intel, Hewlett-Packard to some extent, and others have replaced them. What has in fact happened is that the country has moved to a knowledge economy, and the information being created in private-sector research laboratories has become too valuable to be turned over to a public good. "Before, people were creating physical devices that could be protected in certain ways, through manufacturing knowledge and so forth," he explained, and so the output of research organizations, having a different value, could go into the open literature as "pure public knowledge."

Beginning with the enactment of Bayh-Dole, which Dr. Myers regarded as "a good idea on the whole," there has been a continual privatization of information. Because of the way in which this privatization has taken place, he said, "we do need to worry about the balance of what I call 'the public commons' and what is privately held, [as] there is a creative balance which becomes important." While a great deal of research is being conducted in both government and industry, he said that he was concerned that "a high level of freely appropriable basic knowledge [remains] available to fuel the innovation process," warning that "there has been a trend somewhat away from support of that."

RESEARCH FUNDING CRITICAL TO TRAINING SCIENTISTS

Olwen Huxley, a staff member of the House Committee on Science, asserted that government funding for basic research underwrites most of the education of U.S. Ph.D.s and post-docs. A reduction in such funding would eventually translate to fewer people coming out of labs with the background necessary to go into industry. And technology transfer takes place through people: No innovations could result from the biggest fund of knowledge, even if openly available, without people actually doing the work necessary to transfer a basic idea into a product. "So ultimately we cut our own throats" in the absence of adequate funding for basic research, she said, "but we'll bleed to death over a period of several years, so we may not actually notice it happening." While disavowing powers of clairvoyance, she said that no major increases in such funding are to be anticipated in light of the United States' current level of budgetary deficit, "unless there's some massive sea change that says we're going to go into even more debt to fund basic research."

IS EUROPE MOVING TOWARD MORE BASIC RESEARCH?

Dr. Kuhlmann, commenting from a European perspective on the question of the importance of basic research to innovation, said that discussion there over whether basic research or innovation should receive more focus had moved back and forth. Many programs, at both the national and European levels, have long supported what he called "longer term applied research," which has then stimulated collaboration with industry. Increasingly, however, companies have begun to advocate taking the public-good argument voiced by Dr. Myers more into account. For all the reasons that others had mentioned, knowledge creation and open access to basic advanced knowledge are very important, in particular for innovation. His personal assessment was that Europe is turning in the direction of more basic research, in part because of an increasing perception that the creation of a Europe-wide mechanism to support basic research is needed.

As a result of this sentiment, concrete plans have arisen to establish a European Research Council within the new Framework Program in the coming 1 to 2 years. Although not a copy of the U.S. National Science Foundation (NSF), this new body would be modeled on NSF to some extent. The basic idea is to create in Europe something similar to what the United States has: a huge, single research area with a strong body for funding research.

A EUROPE-WIDE EFFORT IN "FRONTIER RESEARCH"

As a member of a high-level experts group working for the European Commission on this issue, Dr. Kuhlmann has contributed to a report to be released in the days ahead on the potential role of such a European Research Council. This group has argued that such a funding body would achieve the greatest impact if it supports not just basic research but also what is termed "frontier research." As defined in the report, frontier research is collaborative, problem-oriented research into new problems that cut across disciplinary boundaries and might entail not only basic scientific research but also, depending upon the project, engineering, social science, or economics research. A European Research Council is needed to support such groundbreaking work because it would be only through competition for funds on a Europe-wide scale that the best projects and people could be identified.

COST PER PATENT: NATIONAL LABS VS. INDIVIDUALS

Jim Mallos of Heliakon noted that the U.S. Patent Office charges individuals and small businesses thousands of dollars in scheduled fees to process patents on their inventions, while data presented by Dr. VanDevender put at $10 million the per patent cost to the U.S. taxpayer for work at the National Labs. Positing that individuals and small businesses that hold patents were being "fined for doing what we would have paid $10 million at the National Labs to accomplish," he asked whether this does not constitute "a perverse incentive for innovation."

Dr. Myers speculated that the intention of Dr. VanDevender, who had in the meantime departed, was to use that figure as an indicator of productivity. "I couldn't come up with a very good rationale for the unit of analysis," he admitted, "so I would not want to defend his number." An employee of Sandia Labs, commenting from the audience, then said that because a great deal of money is spent there on activities that for a number of reasons "wouldn't possibly be patented," the number in question is unlikely to be reflective of actual productivity. Mr. Mallos, however, reiterated his wonderment that fees of such magnitude are charged to inventors when it is national policy to promote American-owned inventions and patents, whereupon Dr. Myers said his concern was accepted and would be noted.

RISK MANAGEMENT TECHNIQUES FOR POLICY OPTIONS?

Egils Milbergs of the Center for Accelerating Innovation then raised a question about "valuing innovation strategy and innovation policy," which he equated to "valuing what is increasingly an intangible." According to the description offered earlier by Dr. Marburger, university research parks in the United States and abroad seem to share "many of the same thematics: bio, info, nano, and so on." It was his own conviction, Mr. Milbergs said, that not all of these parks or all tech-led economic clusters are "going to win," and that innovation means taking risk, whether that be policy risk, investment risk, career risk, or another sort of risk. His question for all those involved in innovation, therefore, was: Where is the state of the art in what might be called "risk management" that would make possible taking some sensible decisions or creating a sensible portfolio of policy initiatives and investment issues. His motive for asking, he said, lay in his sense that it is enthusiasm and political emotionalism, rather than any serious analysis of true uncertainties and risks that drives many innovation programs.

Mr. Stanley began his response by recalling that during the Reagan Administration, the Department of Commerce, through the Economic Development Administration, conducted a review of what the states were establishing as their strategic technology assets and of how they were employing them. It found that fewer than ten states had either the leadership or the qualifications to make decisions about a number of issues concerning their universities: what goals to set for tech transfer, how to use their assets in collaboration with industry, which options to pursue and which to shun, how to keep their graduate students from leaving the state, and so on. Battelle later became involved in such issues, as did several trade associations.

STATES LEAVING LONG-TERM CONSIDERATIONS ASIDE

Mr. Milbergs had identified an important issue, Mr. Stanley said, one to which no existing formula can be applied and that needs to be reviewed. Luck is generally involved, but very strong leadership is also required at the level of the state governor. As has become clear in recent years, however, many governors are interested in moving into higher office and therefore, when it comes to investing in work for their states, see the "long term" as 4 years rather than 10 or 15.

As for universities, they were busy fighting over technology transfer and intellectual-property issues; or fighting over whether to take warrants or options versus taking money out right away; or fighting with their state legislators for money. "And we're not utilizing the assets of retired businessmen and women that have worked very hard in this area and could be very helpful," he lamented.

In sum, said Mr. Stanley, the United States is a long way from understanding how to develop the various techniques that promote innovation. "And while university parks are good, and clusters are good, and Research Triangle Park is good," he allowed, "I don't think we've gotten the magic potion down yet."

Dr. Myers remarked that, as part of the National Academies study of ATP, he and a colleague had written a subsection about managing portfolios of R&D projects as portfolios of risk. "You not only have to look at the technical risk, you have to look at market risk, and at the interaction of market risk and the technical risk," he pointed out.

ATP, PENTAGON FEDERAL LEADERS IN PROGRAM ASSESSMENT

Dr. Edelheit added that venture capitalists know how to measure risk for the kinds of businesses they invest in, and that there are good examples of industry's best practices in wrestling with the kind of risks it takes on. For government-industry partnerships, such data are to be found in two places. One is in the assessments of ATP, which looks at rewards and risks and at measurements of them, and which he proposed as a good model for those undertaking similar efforts. The second was at the Department of Defense, which does not take a "laissez-faire attitude" toward funding R&D but uses some very clear measurements of risk and reward. He conceded that the latter might be "hard to get" but expressed the opinion that they in fact exist.

PLUSES AND MINUSES OF U.S. POLICY FRAGMENTATION

Larry Rausch of the National Research Foundation said that a conclusion might be drawn from the day's presentations that, in the United States, national innovation policy is in truth a patchwork of different programs. Descriptions of the other countries' more focused innovation policies have raised the following questions: whether the United States needs any more coordinated innovation policy and, if so, what sort of information is needed to guide it or to judge its performance.

Dr. Wessner, praising the question as very interesting, opined that the U.S. system is a patchwork but pointed out that its lack of coherence is both a strength and a weakness. The strength lies in not having a Ministry of Science or a Ministry of Industry that gets it completely wrong. The downside of the United States' distributed system is that "no one's watching the store." As an example of this lack of coherency, he noted that previous analysis by the STEP board first pointed out that the United States cut its R&D budgets for physics, chemistry, and engineering seriously, on a sustained basis, and in real terms between 1993 and 1999 as different agencies responded independently to the end of the cold war.[19] On

---

[19]National Research Council, *Trends in Federal Support of Research and Graduate Education*, Stephen A. Merrill, ed., Washington, D.C.: National Academy Press, 2001.

the other hand, Dr. Wessner stated, when OSTP was fully staffed—which at that moment it was not—there were more people watching the store. That leadership is critical, as is the funding, and the two are related.

"I would argue," Dr. Wessner declared, "that we have some of the best mechanisms in the world that are inadequately fueled." Surprisingly, he observed, the United States was beginning to suffer what he referred to as "the tyranny of small scale": In a $10 trillion or $11 trillion economy, good programs were funded for $10 million, or for $20 million, or, in the case of ATP, for $140 million. "You know it's a good program, but you'd have scale effects that are larger," he added.

III
APPENDIXES

Appendix A

Biographies of Speakers[*]

ALICE H. AMSDEN

Alice H. Amsden is the Barton L. Weller Professor of Political Economy at MIT's Department of Urban Studies and Planning. Professor Amsden's research interests focus on economic and industrial development. She has recently done a research project with the Asian Development Bank Institute on research and development by foreign firms in developing countries. She is currently working with the ADBI on a project on the Indian software industry.

Dr. Amsden recently received the Leontief Prize for Advancing the Frontiers of Economic Thought, awarded by Tufts University's Global Development and Environment Institute Scientific American Top 50 Visionaries. Her selected recent publications include *Beyond Late Development: Taiwan's Upgrading Policies*, coauthored with Wan-wen Chu (MIT Press, June 2003); *The Rise of "the Rest": Challenges to the West from Late-Industrializing Countries* (Oxford University Press, 2001); *The Market Meets its Match: Restructuring the Economies of Eastern Europe*, coauthored with Jacek Kochanowicz and Lance Taylor (Harvard University Press, 1994); and *Asia's Next Giant: South Korea and Late Industrialization* (Oxford, 1989), which won the prize of Best Book in Political Economy in 1992 from the American Political Science Association.

Additionally, Dr. Amsden has served as a consultant with the World Bank, OECD, and various United Nations organizations. She has written extensively on problems of industrial transformation in East Africa, East Asia, and Eastern Europe.

---

[*]As of April 2005.

179

HSIN-SEN CHU

Hsin-Sen Chu is the executive vice president of the Industrial Technology Research Institute (ITRI) in Taiwan. ITRI is a nonprofit R&D organization whose mission is to engage in applied research and technical services to accelerate the industrial development of Taiwan; develop key, compatible, forward-looking technologies to meet industrial needs and strengthen industrial competitiveness; disseminate research results to the industrial sector in a timely and appropriate manner, in accordance with the principles of fairness and openness; foster the technology development of small- and medium-sized businesses; and cultivate industrial technology human resources for the benefit of the nation.

Dr. Chu has served with ITRI since 2001, first as vice president and general director for Energy and Resources Laboratories before assuming his current position in 2004. He has held several R&D related positions, including serving as director of the High Efficiency Energy Technology Research Center at National Chiao Tung University from 2000 to 2001; director of the Mechanical Manufacturing & Heat Flow Research Center at National Chiao Tung University from 1999 to 2001; chief of staff at National Chiao Tung University from 1995 to 1998; vice dean of the College of Engineering at National Chiao Tung University from 1992 to 1993; chairman of the Department of Mechanical Engineering at National Chiao Tung University from 1991 to 1995; professor in the Department of Mechanical Engineering at National Chiao Tung University from 1989 to the present; visiting scholar at the University of California, Berkeley from 1985 to 1986; and associate professor in the Department of Mechanical Engineering at National Chiao Tung University from 1984 to 1989.

Dr. Chu has received many research awards in recognition of his contributions, including the Distinguished Research Award from the National Science Council of Taiwan in 1999; the Distinguished Engineering Professor Award from the Chinese Society of Mechanical Engineering and the Chinese Society of Engineers, R.O.C in 1998; the Excellent Young Engineer Award from the Chinese Society of Mechanical Engineering, R.O.C. in 1991; and the Excellent Research Award from the National Science Council, R.O.C. 1988-1993. He was named in the 1997 Marquis Who's Who in the World and has published 60 technical journal papers, 44 conference papers, and 30 technical reports in his career.

Dr. Chu received his B.S. in 1974, M.S. in 1977, and Ph.D. in 1982 from National Cheng Kung University, Taiwan, R.O.C. He was named an Honorary Member of Phi Tau Phi in 1977 at National Cheng Kung University.

CARL J. DAHLMAN

Carl J. Dahlman, is the Luce Professor of International Affairs and Information Technology at the Edmund A. Walsh School of Foreign Service at Georgetown University. He joined Georgetown in January 2005 after more than 25 years of distinguished service at the World Bank.

At Georgetown, Dr. Dahlman's research and teaching will explore how rapid advances in science, technology and information are affecting the growth prospects of nations and influencing trade, investment, innovation, education and economic relations in an increasingly globalizing world. At the World Bank Dr. Dahlman served as senior advisor to the World Bank Institute. In this role he managed the Knowledge for Development (K4D) program, an initiative providing training on the strategic use of knowledge for economic and social development to business leaders and policy makers in developing countries. Prior to developing the K4D program, Dr. Dahlman served as staff director of the 1998-1999 World Development Report, *Knowledge for Development*. In addition, he was the Bank's Resident Representative and Financial Sector Leader in Mexico from 1994 to 1997, years during which the country coped with one of the biggest financial crises in its history. Before his position in Mexico, Dr. Dahlman had led divisions in the Bank's Private Sector Development, and Industry and Energy Departments. He has also conducted extensive analytical work in major developing countries including Argentina, Brazil, Chile, Mexico, Russia, Turkey, India, Pakistan, China, Korea, Malaysia, Philippines, Thailand, and Vietnam.

Dr. Dahlman's publications include *India and the Knowledge Economy: Leveraging Strengths and Opportunities* (2005), *China and the Knowledge Economy: Seizing the 21st Century* (2001), and *Korea and the Knowledge-Based Economy: Making the Transition* (2000). He is currently finishing a knowledge economy study on Mexico, working on a book on the challenge of the knowledge economy for education and training in China, and collaborating with research teams in Finland, Japan, and Korea to produce books on each country's innovation and development strategies.

Dr. Dahlman earned a B.A. magna cum laude in international relations from Princeton University and a Ph.D. in economics from Yale University. He has also taught courses at Columbia University's School of International and Public Affairs.

LEWIS S. EDELHEIT

Lewis S. (Lonnie) Edelheit retired in 2001 from his position as senior vice president, Corporate R&D, General Electric Company. Under his leadership, GE introduced numerous new leadership products, including digital X-ray and advanced ultrasound medical imagers, high-efficiency turbines for power generation, advanced lighting and electronics-based appliances and weatherable plastics to name a few. Other highlights of his tenure include significant advances in high-technology services and Internet applications and Corporate R&D's leadership of the Design for Six Sigma quality and e-Engineering initiatives throughout the GE businesses. Also under his leadership, Corporate R&D vastly expanded its global resources with the development of new technology centers in Bangalore, India, and Shanghai, China. Dr. Edelheit began his professional career in 1969 as a physicist at the GE R&D Center, where he made significant contributions to fast-scan, "fan-beam" computed tomography x-ray systems.

In 1976, Dr. Edelheit transferred to GE Medical Systems in Milwaukee, Wisconsin, where he rose to such positions as general manager of engineering and general manager of the Computed Tomography Programs Department, where he held marketing and profit-and-loss responsibility for GE's worldwide computed tomography scanning business. In 1986, Dr. Edelheit left GE to become president and CEO of Quantum Medical Systems, a venture capital-backed company that pioneered color flow ultrasound for vascular imaging. He continued in that position after Quantum was acquired by the Siemens Corporation. In 1991 he returned to GE and assumed leadership of Corporate R&D.

Dr. Edelheit earned a B.S. degree in engineering physics and an M.S. degree and Ph.D. in physics from the University of Illinois. In 1995, he received the University of Illinois at Urbana-Champaign College of Engineering Alumni Award for Distinguished Service. Dr. Edelheit is a member of the National Academy of Engineering, the Industrial Research Institute (named as the 2003 Medalist), and a fellow of the American Physical Society, which selected him as the recipient of the 2001 George E. Pake prize. He is on advisory boards of the Physics Department and Bio Engineering Department of the University of Washington and of the Harvard Medical and Beth Israel Deaconess Shapiro Research and Education Institute in Boston. He is also on the board of directors of two public corporations, Silicon Graphics, and Sonic Innovation, two private corporate boards, and is chair of the Laboratory Advisory Committee of the Pacific Northwest National Laboratory.

KENNETH FLAMM

Kenneth Flamm is the Dean Rusk Professor of International Affairs at the Lyndon B. Johnson School of Public Affairs at the University of Texas at Austin.

Dr. Flamm is a 1973 honors graduate of Stanford University and received a Ph.D. in economics from M.I.T. in 1979. From 1993 to 1995, Dr. Flamm served as principal deputy assistant secretary of Defense for Economic Security and special assistant to the Deputy Secretary of Defense for Dual Use Technology Policy. He was awarded the Department's Distinguished Public Service Medal in 1995 by Defense Secretary William J. Perry. Prior to his service at the Defense Department, he spent eleven years as a senior fellow in the Foreign Policy Studies Program at the Brookings Institution.

Dr. Flamm has been a professor of economics at the Instituto Tecnológico A. de México in Mexico City, the University of Massachusetts, and George Washington University. He has also been an adviser to the Director General of Income Policy in the Mexican Ministry of Finance and a consultant to the Organization for Economic Cooperation and Development, the World Bank, the National Academy of Sciences, the Latin American Economic System, the U.S. Department of Defense, the U.S. Department of Justice, the U.S Agency for International Development, and the Office of Technology Assessment of the U.S. Congress.

Among Dr. Flamm's publications are *Mismanaged Trade? Strategic Policy and the Semiconductor Industry* (1996), *Changing the Rules: Technological Change, International Competition, and Regulation in Communications* (ed., with Robert Crandell, 1989), *Creating the Computer* (1988), *and Targeting the Computer* (1987). He is currently working on an analytical study of the post-Cold War defense industrial base.

Dr. Flamm, an expert on international trade and high-technology industry, teaches classes in microeconomic theory, international trade, and defense economics.

THOMAS R. HOWELL

Thomas R. Howell is a partner with the Washington, D.C., law office of Dewey Ballantine LLP, where he specializes in international trade matters. Since 1981 he has written extensively on industrial, trade, and technology polices in countries outside of the United States. In 2003, he prepared a report for the Semiconductor Industry Association, *China's Emerging Semiconductor Industry: The Impact of China's Preferential Value-Added Tax on Current Investment Trends.*

DAVID K. KAHANER

David K. Kahaner is the founding director of the Asian Technology Information Program (ATIP), which was established in 1994. He was formerly the associate director of the U.S. Office of Naval Research Asia. He also spent more than twenty years at the U.S. National Institute of Standards and Technology (formerly the National Bureau of Standards) and a dozen years at the Los Alamos National Laboratory.

Dr. Kahaner has been examining information rich technologies in Asia for many years. His analyses are circulated worldwide to thousands in industry, government, and academia. They are reprinted in many journals as well as often quoted in major news media, and he consults and lectures frequently on those topics both in and outside of the region. In 1993, he was awarded the title of "Mr. Asia" by *Computerworld*. One of his goals is to help Westerners understand opportunities and issues associated with science-based activities in the Asian region.

Dr. Kahaner has been the Asian chair for a variety of international conferences and chair of the International Organizing Committees for the conference series HPC-Asia, which has been held seven times in as many different Asian cities since 1995. Separately, Dr. Kahaner also has many years of research experience in scientific computing and applications. Many of the applications he and his groups developed are used in scientific computing centers worldwide, and he has received several national awards for this work. He is the author of two well known textbooks and more than 50 refereed research papers. He has edited a column on scientific applications of computers, and has held numerous journal editorial and associate editorial positions. He has had visiting professorships at major universities in the U.S, Austria, Italy, and Switzerland, where he has taken extensive sabbaticals and still retains significant associations.

BRADLEY KNOX

Bradley Knox is the chief counsel for the U.S. House of Representatives Committee on Small Business, serving in the number-three management and policy role under the chief of staff and deputy chief of staff/policy director. Mr. Knox joined the committee in early 2003 as oversight counsel, responsible for oversight of federal agencies in government procurement and other matters, and served as staff manager of the chairman's initiative to revitalize the U.S. manufacturing base.

During the 1990s, Mr. Knox was an Air Force JAG officer for six years and then in active reserves, taught at the Air Force JAG School on the legal aspects of information warfare and homeland security for two years. He spent his last two years as a reservist assigned as legal counsel to the Joint Task Force on Global Network Attack.

Mr. Knox began his professional career with Shell Oil as a computer programmer and analyst. He was cofounder and president of DigiTech System Solutions, an info-tech services company headquartered in Montgomery, Alabama. He is cofounder and vice president of Knox Global Network, a business-to-business and business-to-consumer e-commerce franchise.

Mr. Knox was honored as one of the Top 40 business leaders under the age of 40 in Central Alabama, and DigiTech was named one of Montgomery's Emerging Businesses in 2002. He earned a B.S. in information systems at Oral Roberts University and a J.D. at Regent University.

HEIKKI KOTILAINEN

Heikki Kotilainen is the deputy director general of Tekes, the Finnish National Technology Agency. He has a degree of Dr. of Techn. in mechanical engineering from Helsinki University of Technology. He has studied in Germany and Austria and worked in industry, research institutes, and public administration. For over 16 years he has worked in financing R&D in industry and universities, participated in formulating government technology and innovation policy, coordinated national technology programs and international cooperation, and served as a member of many domestic and international (EU, Nordic) technology and innovation committees and boards. Serving in the United States from 1993 to 1995 gave him a very thorough insight into the research and high-tech industry developments (MIT, Harvard, Boston Route 128). From 2000 to 2003, he was the head of the EUREKA Secretariat in Brussels, running the industrial cooperation platform in Europe among 31 countries and participating in the ERA discussions from the innovation point of view. In recent years, his duty has been the strategic planning of the National Technology Agency, Tekes, as the deputy director general. Part of his work is to look for the best practices in leading technology and innovation policy countries. Lately, he has been lecturing in many countries about technology and innovation policy and management.

STEFAN KUHLMANN

Stefan Kuhlmann, a political scientist, is director of the Fraunhofer Institute for Systems Innovation Research (ISI), Karlsruhe, Germany. He is also professor of Innovation Policy Analysis at the Copernicus Institute, Innovation Studies Group, University of Utrecht, the Netherlands. He has more than 15 years background in the assessment and analysis of research and innovation policies.

Dr. Kuhlmann has published widely in the field of research and innovation policy studies. He is an associate editor of the *International Journal of Foresight and Innovation Policy (IJFIP)*, and serves on the editorial advisory board of *Evaluation: The International Journal of Theory, Research and Practice*, edited in association with The Tavistock Institute, London, UK, since 2000. Recent publications include *Changing Governance of Research and Technology Policy: the European Research Area*, Cheltenham (E. Elgar) 2003 (coedited with J. Edler and M. Behrens); *Learning from Science and Technology Policy Evaluation: Experiences from the United States and Europe*, Cheltenham (E. Elgar) 2003 (co-edited with Ph. Shapira); "Evaluation of research and innovation policies: a discussion of trends with examples from Germany," *International Journal of Technology Management* (26, Nos. 2/3/4, 2003); and "Governance of Innovation Policy in Europe: Three Scenarios," *Research Policy* (30, June 2001).

Dr. Kuhlmann is a member of many professional or academic associations in the area of research and innovation policy analysis, including the Executive Committee European Network of Excellence PRIME (Policies for Research and Innovation on the Move towards the European Research Area); the European Commission's High-level Expert Group on "Maximising the wider benefits of competitive basic research funding at European level" (Directorate General Research); the European RTD Evaluation Network of the European Commission, Directorate General Research; the Netherlands Graduate School of Science, Technology and Modern Culture (WTMC); the steering committee of the Six Countries Programme—The International Innovation Network; the Scientific Advisory Committee of the Helsinki Institute of Science and Technology Studies (HIST);

and the Board of the German Evaluation Association (DeGEval), on which he
served from 1997 to 2001.

JOHN H. MARBURGER

John H. Marburger, III, science adviser to the President and director of the
Office of Science and Technology Policy, was born on Staten Island, N.Y., grew
up in Maryland near Washington, D.C., and attended Princeton University (B.A.,
physics, 1962) and Stanford University (Ph.D., applied physics, 1967). Before
his appointment in the Executive Office of the President, he served as director
of Brookhaven National Laboratory from 1998, and as the third president of the
State University of New York at Stony Brook (1980-1994). He came to Long
Island in 1980 from the University of Southern California where he had been a
professor of physics and electrical engineering, serving as Physics Department
chairman and dean of the College of Letters, Arts and Sciences in the 1970s. In
the fall of 1994 he returned to the faculty at Stony Brook, teaching and doing
research in optical science as a university professor. Three years later he became
president of Brookhaven Science Associates, a partnership between the university
and Battelle Memorial Institute that competed for and won the contract to operate
Brookhaven National Laboratory.

While at the University of Southern California, Dr. Marburger contributed to
the rapidly growing field of nonlinear optics, a subject created by the invention
of the laser in 1960. He developed theory for various laser phenomena and was
a cofounder of the University of Southern California's Center for Laser Studies.
His teaching activities included "Frontiers of Electronics," a series of educational
programs on CBS television.

Dr. Marburger's presidency at Stony Brook coincided with the opening and
growth of University Hospital and the development of the biological sciences as a
major strength of the university. During the 1980s, federally sponsored scientific
research at Stony Brook grew to exceed that of any other public university in
the northeastern United States. During his presidency, Dr. Marburger served on
numerous boards and committees, including chair of the governor's commission
on the Shoreham Nuclear Power facility, and chair of the 80 campus Universities
Research Association, which operates Fermi National Accelerator Laboratory
near Chicago. He served as a trustee of Princeton University and many other
organizations. He also chaired the highly successful 1991-92 Long Island United
Way campaign.

As a public spirited scientist-administrator, Dr. Marburger has served local,
state, and federal governments in a variety of capacities. He is credited with
bringing an open, reasoned approach to contentious issues where science inter-
sects with the needs and concerns of society. His strong leadership of Brookhaven
National Laboratory following a series of environmental and management crises

is widely acknowledged to have won back the confidence and support of the community while preserving the laboratory's record of outstanding science.

MARK B. MYERS

Mark B. Myers is visiting executive professor in the Management Department at the Wharton Business School, the University of Pennsylvania. His research interests include identifying emerging markets and technologies to enable growth in new and existing companies with special emphases on technology identification and selection, product development, and technology competencies. Dr. Myers serves on the Science, Technology and Economic Policy Board of the National Research Council and cochaired, with Richard Levin, the president of Yale, the National Research Council's study of "Intellectual Property in the Knowledge Based Economy."

Dr. Myers retired from the Xerox Corporation at the beginning of 2000, after a 36-year career in its research and development organizations. He was the senior vice president in charge of corporate research, advanced development, systems architecture, and corporate engineering from 1992 to 2000. His responsibilities included the corporate research centers, PARC in Palo Alto, California; Webster Center for Research & Technology near Rochester, New York; Xerox Research Centre of Canada, Mississauga, Ontario; and the Xerox Research Centre of Europe in Cambridge, UK, and Grenoble, France. During this period he was a member of the senior management committee in charge of the strategic direction setting of the company.

Dr. Myers is chair of the board of trustees of Earlham College and has held visiting faculty positions at the University of Rochester and at Stanford University. He holds a bachelor's degree from Earlham College and a doctorate from Pennsylvania State University.

PETER J. NICHOLSON

Peter J. Nicholson is Deputy Chief of Staff–Policy, Office of the Prime Minister of Canada. A native of Halifax, Nova Scotia, he holds a B.Sc. and M.Sc. in physics from Dalhousie University and a Ph.D. in operations research from Stanford University, as well as honorary doctorates from Acadia University, Dalhousie and the Université du Québec (INRS). After post-doctoral work in France, Dr. Nicholson joined the Computer Science Department at the University of Minnesota in 1969 where he taught four years before joining the Government of Canada in 1973. There he served in a senior policy advisory role in the Departments of Urban Affairs, Transport, and Regional Economic Expansion.

In 1978, Dr. Nicholson left Ottawa and was elected to the Legislature of the Province of Nova Scotia. At the time, he became associated with H. B. Nickerson & Sons, a major fisheries company, and eventually left the Legislature to devote full-time to the company as a vice president. In 1982, he joined the Taskforce on Atlantic Fisheries established by the federal government to restructure the industry, which had been financially devastated by the 1981-1982 recession. In 1984 Dr. Nicholson joined The Bank of Nova Scotia in Toronto where he was senior vice president, advising the chair of the bank on a broad range of strategic issues, including in particular the resolution of the Latin American debt crisis in the late 1980s.

Between March 1994 and September 1995, Dr. Nicholson was Clifford Clark Visiting Economist in the federal Department of Finance under the government's executive interchange program. This is a senior advisory position to Canada's Minister and Deputy Minister of Finance. From September 1995 to June 2002, he was chief strategy officer of BCE, Inc., Canada's largest telecommunications company. Between June 2002 and July 2003, he was special adviser to the Secretary-General of the Organization for Economic Co-operation and Development in Paris. He joined the Prime Minister's Office in December 2003. Dr. Nicholson is a member of the Order of Canada, in recognition of his contribution to business through both the public and private sectors.

HIDEO SHINDO

Hideo Shindo is the chief representative of New Energy and Industrial Technology Development Organization (NEDO) in its Washington, D.C. office, where he has served since July 2004. NEDO is a Japanese semigovernment nonprofit organization whose objective is to facilitate R&D activities in industries in Japan, in close cooperation with Japanese government agencies such as METI (Ministry of Economy, Trade, and Industry). Mr. Shindo entered METI in 1986 and dealt with a variety of policy issues such as technology development, international trade and investment, and codes and standards. In 2003 he was transferred to NEDO headquarters and assisted in its restructuring as an incorporated administrative agency.

WILLIAM J. SPENCER

William J. Spencer was named chairman emeritus of the SEMATECH Board in November 2000 after serving as chairman of the SEMATECH and International SEMATECH Boards since July 1996. He came to SEMATECH in October 1990 as president and chief executive officer. He continued to serve as president until January 1997 and as CEO until November 1997. During this time, SEMATECH became totally privately funded and expanded to include non-U.S. members. Many analysts credit SEMATECH with making a major contribution to the recovery of the U.S. semiconductor industry in the 1990s and its recapture of market share.

Dr. Spencer has held key research positions at Xerox Corporation, Bell Laboratories, and Sandia National Laboratories. Before joining SEMATECH in October 1990, he was group vice president and senior technical officer at Xerox Corporation in Stamford, Connecticut, from 1986 to 1990. He established new research centers in Europe and developed a plan for Xerox retaining ownership in spinout companies from research. Prior to joining the Xerox Palo Alto Research Center (PARC) as manager of the Integrated Circuit Laboratory in 1981 and as the center manager of PARC in 1982 to 1986, Dr. Spencer served as director of Systems Development from 1978 to 1981 at Sandia National Laboratories in Livermore, and director of Microelectronics at Sandia National Laboratories in Albuquerque from 1973 to 1978, where he developed a silicon processing facility for Department of Energy needs. He began his career in 1959 at Bell Laboratories.

Dr. Spencer received the Regents Meritorious Service Medal from the University of New Mexico in 1981; the C. B. Sawyer Award for contribution to "The Theory and Development of Piezoelectric Devices" in 1972; and a Citation for Achievement from William Jewell College in 1969, where he also received an doctor of science degree in 1990. He is a member of the National Academy of Engineering, a Fellow of IEEE, and serves on numerous advisory groups and boards. He was the Regents Professor at the University of California in the spring of 1998. He has been a visiting professor at the University of California at Berkeley School of Engineering and the Haas School of Business since the fall of 1998. He is a research professor of medicine at the University of New Mexico.

Dr. Spencer received an A.B. degree from William Jewell College in Liberty, Missouri, an M.S. degree in mathematics and a Ph.D. in physics from Kansas State University.

MARC G. STANLEY

Marc G. Stanley currently serves as director of the Advanced Technology Program (ATP) at the National Institute of Standards and Technology (NIST). Mr. Stanley served as the associate director for the program from 1993 to 2001.

Before coming to NIST, Mr. Stanley was the Associate Deputy Secretary of the U.S. Department of Commerce by Presidential appointment. He has served as a senior policy advisor to NIST directors, as a consultant to the Department of Commerce's Technology Administration, and as Assistant Secretary for Congressional and Intergovernmental Affairs at the Department of Commerce.

Mr. Stanley earned a B.A. from George Washington University and a bachelor of law degree from the University of Baltimore.

J. PACE VANDEVENDER

J. Pace VanDevender is vice president of Science & Technology and Partnerships and chief technology officer at Sandia National Laboratories. He is also chief scientific officer for the nuclear weapons program at the laboratory and is accountable for Sandia's partnership program with industry and the university community.

Dr. VanDevender leads and/or manages the research, development and engineering in nanosciences, materials and process sciences, microelectronics/microsystems and optoelectronics, advanced manufacturing, computational sciences, modeling and simulation science, and high-energy density physics.

In 1974, he joined Sandia as a member of the technical staff. Dr. VanDevender became division supervisor in 1978, manager of the Fusion Department in 1982, and director of Pulsed Power Sciences in 1984. In the latter position, which he held into the 1990s, he oversaw Sandia's pulsed power R&D, inertial confinement fusion research, nuclear weapons effects simulations, directed energy weapons R&D, and commercial applications of pulsed power.

In 1992, Dr. VanDevender received DoE's prestigious E.O. Lawrence Award, "for his outstanding contributions to the generation of pulsed power," becoming only the fifth Sandian to receive that award.

In 2003, he was named a fellow of the American Association for the Advancement of Science. He is also a fellow of the American Physical Society and a senior member of the Institute of Electrical and Electronics Engineers.

As director of executive staff at Sandia National Laboratories, Dr. VanDevender was in charge of governance initiatives, strategic planning, government relations, quality assurance, laboratory assessment, issues management, and the support of the Sandia Executives.

Before becoming the executive staff director, Dr. VanDevender was the chief information officer and was responsible for the information infrastructure, for cyber security, and for making information technology a compelling advantage for doing science, engineering, and manufacturing. He has also been the director of the Strategic Sciences Center, National Industrial Alliance Center, Corporate Communications Center, and Pulsed Power Sciences Center.

Dr. VanDevender earned a Ph.D. in physics from the Imperial College of Science and Technology, University of London, England in 1974, where he was a Marshall Scholar; an M.A. in physics from Dartmouth College in 1971; and a B.A. in physics from Vanderbilt University in 1969.

CHARLES W. WESSNER

Charles W. Wessner is an internationally recognized expert on innovation policy, serves as the National Research Council director for *Comparative Innovation Policy: Best Practice in National Technology Programs*, and is responsible for organizing this conference. He directs several other studies at the National

Academies, notably *Measuring and Sustaining the New Economy* and *Capitalizing on Science Technology, and Innovation: An Assessment of the Small Business Innovation Research Program.*

Dr. Wessner's expertise centers on public-private partnerships, early-stage financing for new firms, university-industry partnerships, and high-tech clusters, as well as the special needs and benefits of high-technology industry. Dr. Wessner is regularly consulted by congressional staff and testifies to the Congress and major national commissions, acts as an advisor to agencies of the U.S. Government, international organizations such as the Organization for Economic Co-operation and Development (OECD), and leading technology agencies in Europe and Asia, and lectures at major universities in the United States and abroad. He previously worked in the U.S. Treasury Department, served overseas with the OECD and the U.S. State Department, and directed the Office of International Technology Policy at the Department of Commerce.

Dr. Wessner earned an M.A., M.A.L.D., and Ph.D. from the Fletcher School of Law and Diplomacy.

Appendix B

Participants List*

Alice H. Amsden
Massachussets Institute of Technology

Tim Angus
Industry Canada

Martin Apple
Council of Scientific Society
    Presidents

Clara Asmail
National Institute of Standards and
    Technology

David Attis
Council on Competitiveness

Gary Bachula
Internet2

Robert Bell
National Science Foundation

Tabitha Benney
National Research Council

Arpad Bergh
Optoelectronics Industry
    Development Association

Richard Bissell
National Research Council

Peter Blair
National Research Council

Rochelle Blaustein
National Institutes of Health

Robert Boege
ASTRA

Jeff Bond
Association for Manufacturing
    Technology

Speakers in italics.

William Bonvillian
Office of Senator Lieberman

Mark Boroush
U.S. Department of Commerce

Edsel Brown
U.S. Small Business Administration

Alphonse Buccino

Amy Burke
Semiconductor Industry Association

Dan Byers
House Committee on Science

Simon Cavenett
Detecon

Bill Chadwick
U.S. International Trade Commission

Connie Chang
National Institute of Standards and
    Technology

John Chester

Hsin-Sen Chu
Industrial Technology Research
    Institute (ITRI), Taiwan

McAlister Clabaugh
National Research Council

Major Clark
U.S. Small Business Administration

William Colglazier
National Research Council

Eileen Collins
Rutgers University

Mildred Cooper

Ronald Cooper
U.S. Small Business Administration

Carol Corrado
Federal Reserve Board

Michael Czinkota
Georgetown University

Carl J. Dahlman
Georgetown University and
    World Bank, retired

Brian Darmody
University of Maryland

Robert Davidson
Canada Foundation for Innovation

Patrick Dennis
National Science Foundation

Doug Devereaux
U.S. Department of Commerce

Papan Devnani

Bea Droke
U.S. Food and Drug Administration

Travis Earles
National Cancer Institute

Lewis S. Edelheit
General Electric, retired

Giorgio Einaudi
Embassy of Italy

Cerise Elliott
National Institutes of Health

Sara Farley
Rockefeller Foundation

Lenka Fedorkova
National Institutes of Health

Brenda Fisher
U.S. Department of Commerce

Kenneth Flamm
University of Texas at Austin

Pamela Flattau
Institute for Defense Analysis

Quindi Franco
SRI International

Susan Fratkin
Fratkin Associates

Mariko Fukumura
Washington CORE

Edward Furtek
Univeristy of California, San Diego

James Gallup
U.S. Environmental Protection
    Agency

Eric Garduño
International Intellectual Property
    Institute

Robin Gaster
North Atlantic Research

Vinod Goel
The World Bank

Rich Golaszewski
GRA Inc.

Randy Goodall
SEMATECH

Jo Anne Goodnight
National Institutes of Health

Margaret Grabb
National Institutes of Health

James Hague
House Committee on Science

Gerald Hane
Globalvation

Angela Hardin
Inside Energy

John Hardin
North Carolina Board of S&T

Meg Hardon
Infineon Technologies

David Hart
George Mason University

Christopher Hayter
NACFAM

Bill Hendrickson
Issues in Science and Technology

Bart Hendrickx
Embassy of Belgium

Robert Hershey

Mat Heyman
National Institute of Standards and
    Technology

Chris Hill
George Mason University

Derek Hill
National Science Foundation

Pamela Houghtaling
National Institute of Standards and
    Technology

Betsy Houston
Federation of Materials Societies

Thomas R. Howell
Dewey Ballantine

Virginia Hughes
Government Accountability Office

Hiroshi Ikukawa
Embassy of Japan

Ken Jacobson

Ken Jarboe
Athena Alliance

Al Johnson
Corning

George Johnson
National Cancer Institute

Richard Johnson
Arnold & Porter

Trevor Jones
BIOMEC

Teresa Jones
NACFAM Weekly

Gretchen Jordan
Sandia National Laboratories

James Kadtke
Office of Senator Warner

David K. Kahaner
Asian Technology Information
    Program

Glenn Kendall
Canadian Biotechnology Secretariat

Emily Kennedy
Embassy of Australia

Chris King
House Committee on Science

Kathleen Kingscott
International Business Machines

Bradley Knox
House Committee on Small Business

Kei Koizumi
American Association for the
    Advancement of Science

Heikki Kotilainen
Tekes, Finland

Charlotte Kuh
National Research Council

Stefan Kuhlmann
Fraunhofer ISI, Germany

Jean-Philippe Lagrange
Embassy of France

Jerri Laine
Tekes

Bhavya Lal
Abt Associates

Jean-Jacques Lawrence
Embassy of France

Emmanuel Le Perru
Anvar

Rolf Lehming
National Science Foundation

Marc LePage
Genome Canada

Nanette Levinson
American University

Ellen Levy
Stanford University

Sira Maliphol
George Washington University

Jim Mallos
Heliakon

Christopher Mapes
U.S. International Trade Commission

John H. Marburger
White House Office of Science and
*   Technology Policy*

Marie Marcoux

Cheryl Marks
National Cancer Institute

Lluana McCann
National Science Foundation

Kari McCarron
American Association for the
    Advancement of Science

Richard McCormack
Manufacturing & Technology News

Bruce McWilliams
George Washington University

Natalia Melcer
American Chemical Society

Stephen A. Merrill
National Research Council

Evan Michelson
George Washington University

Egils Milbergs
Center for Accelerating Innovation

Barbara Mittleman
National Institutes of Health

Mary Ellen Mogee
SRI International

Sushanta Mohapatra
SRI International

David T. Morgenthaler
Morgenthaler Ventures

William Morin
Applied Materials

Francisco Moris
National Science Foundation

Massoud Moussavi
The World Bank

Kazue Muroi
Washington Core

Mark B. Myers
The Wharton School

Kesh Narayanan
National Science Foundation

Pauline Newman
Court of Appeals–Federal Circuit

Susan Nichols
U.S. Army

Peter J. Nicholson
Office of the Prime Minister, Canada

Christopher Niebylski
George Washington University

Robert Norwood
National Science Foundation

Kim Olsen
U.S. International Trade Commission

Emmanuel Onyiriuka
Corning

Paul op den Brouw
Embassy of the Royal Netherlands

Jennifer Ozawa
SRI International

Jongwon Park
SRI International

Luciano Parodi
Embassy of Chile

Lori Perine
American Forest & Paper Association

Pierre Perrolle
National Science Foundation

David Peyton
National Association of Manufacturers

Anthony Pitagno
American Chemical Society

Jean Pomeroy
National Science Foundation

Mike Quear
House Committee on Science

Lisette Ramcharan
Embassy of Canada

Samuel Rankin
American Mathematical Society

Noel Raufaste
National Institute of Standards and
    Technology

Larry Rausch
National Science Foundation

Kalpanna Reddy
U.S. Department of Agriculture

Leslie Reissner
Embassy of Canada

Catherine Renault
RTI International

Volker Rieke
Embassy of Germany

Barrett Ripin
Department of State

Deborah Rudolph
IEEE-USA

Luis Salicrup
National Institutes of Health

Ted Schlie
Lehigh University

Theodore Schmitt
National Research Council

Gregory Schuckman
University of Central Florida

Craig Schultz
National Research Council

Lyle Schwartz

Avery Sen
National Oceanic and Atmospheric
    Administration

Christal Sheppard
House Committee on Science

Chuck Sherman
National Institutes of Health

Hideo Shindo
New Energy and Industrial
    Technology Development
    Organization (NEDO), Japan

Kathleen Shino
National Institutes of Health

Stephanie Shipp
National Institute of Standards and
    Technology

Sujai Shivakumar
National Research Council

Victor Shulepov
Embassy of Ukraine

Robert Sienkiewicz
National Institute of Standards and
    Technology

Mark Skinner
SSTI

Lana Skirboll
National Institutes of Health

Amanda Slocum
National Science Foundation

Stan Sokul
White House Office of Science and
    Technology Policy

Susannah Spellman
George Washington University

William J. Spencer
SEMATECH, retired

Marc G. Stanley
National Institute of Standards and
    Technology

Deborah Stine
National Research Council

Miron Straf
National Research Council

Istvan Takacs
Embassy of Hungary

Greg Tassey
National Institute of Standards and
    Technology

Margaret Thompson
U.S. Department of Energy

Dawn Trembath
North Carolina Board of Science and
    Technology

David Trinkle
Office of Management and Budget

James Turner
House Committee on Science

J. Pace VanDevender
Sandia National Laboratories

David Waechter
University of North Carolina

Andrew Wang
National Institute of Standards and
    Technology

Al Watkins
The World Bank

Philip Webre
Congressional Budget Office

Charles W. Wessner
National Research Council

Jeffrey Williams
George Washington University

Shane Williamson
Industry Canada

Alan Wm. Wolff
Dewey Ballantine

Patricia Wrightson
National Research Council

Falan Yinug
U.S. International Trade Commission

Obaid Younossi
RAND

Nick Zayed
U.S. Department of State

Appendix C

Bibliography

Alic, John A., Lewis M. Branscomb, Harvey Brooks, Ashton B. Carter, and Gerald L. Epstein. 1992. *Beyond Spin-off: Military and Commercial Technologies in a Changing World*. Boston, MA: Harvard Business School Press.

Amsden, Alice H. 2001. *The Rise of "the Rest": Challenges to the West from Late-industrializing Economies*. Oxford, UK: Oxford University Press.

Amsden, Alice H. and Wan-wen Chu. 2003. *Beyond Late Development: Taiwan's Upgrading Policies*. Cambridge, MA: The MIT Press.

Amsden, Alice H., Ted Tschang, and Akira Goto. 2001. "Do Foreign Companies Conduct R&D in Developing Countries?" Tokyo, Japan: ADB Institute.

Archibugi, Danielle, Jeremy Howells, and Jonathan Michie, eds. 1999. *Innovation Policy and the Global Economy*. Cambridge, UK: Cambridge University Press.

Asheim, Bjorn T. et al, eds. 2003. *Regional Innovation Policy for Small-medium Enterprises*. Cheltenham, UK and Northampton, MA: Edward Elgar.

Athreye, Suma S. 2000. "Technology Policy and Innovation: The Role of Competition Between Firms." In Pedro Conceicao et al., eds. *Science, Technology, and Innovation Policy: Opportunities and Challenges for the Knowledge Economy*. Westport, CT, and London, UK: Quorum Books.

Audretsch, David B. 2006. *The Entrepreneurial Society*, Oxford, UK: Oxford University Press.

Audretsch, David B., Heike Grimm, and Charles W. Wessner. 2005. *Local Heroes in the Global Village: Globalization and the New Entrepreneurship Policies*. New York, NY: Springer.

Audretsch, David B., ed. 1998. *Industrial Policy and Competitive Advantage*, Volumes 1 and 2. Cheltenham, UK: Edward Elgar.

Audretsch, D. B., B. Bozeman, K. L. Combs, M. P. Feldman, A. N. Link, D. S. Siegel, P. Stephan, G. Tassey, and C. Wessner. 2002. "The Economics of Science and Technology." *Journal of Technology Transfer* 27:155-203.

Baldwin, John Russel, and Peter Hanel. 2003. *Innovation and Knowledge Creation in an Open Economy: Canadian Industry and International Implications*. Cambridge, UK: Cambridge University Press.

Balzat, Markus and Andreas Pyka. 2006. "Mapping National Innovation Systems in the OECD Area." *International Journal of Technology and Globalisation* 2(1-2):158-176.

Bartzokas, Anthony and Morris Teubal. 2002. "The Political Economy of Innovation Policy Implementation in Developing Countries." *Economics of Innovation and New Technology* 11(4-5).

Bhidé, Amar. 2006. "Venturesome Consumption, Innovation and Globalization." Paper presented at the Centre on Capitalism & Society and CESifo Venice Summer Institute 2006, "Perspectives on the Performance of the Continent's Economies," 21-22 July 2006. Held at Venice International University. San Servolo, Italy.

Biegelbauer, Peter S. and Susana Borras, eds. 2003. *Innovation Policies in Europe and the U.S.: The New Agenda.* Aldershot, UK: Ashgate.

Blomström, Magnus, Ari Kokko, and Fredrik Sjöholm. 2002. "Growth & Innovation Policies for a Knowledge Economy: Experiences from Finland, Sweden, & Singapore." EIJS Working Paper, Series No. 156.

Borras, Susana. 2003. *The Innovation Policy of the European Union: From Government to Governance.* Cheltenham, UK: Edward Elgar.

Borrus, Michael and Jay Stowsky. 2000. "Technology Policy and Economic Growth." In Charles Edquist and Maureen McKelvey, eds. *Systems of Innovation: Growth, Competitiveness and Employment, Vol. 2.* Cheltenham, UK and Northampton, MA: Edward Elgar.

Branscomb, Lewis M. and Philip E. Auerswald. 2002. *Between Invention and Innovation: An Analysis of Funding for Early-Stage Technology Development.* NIST GCR 02–841. Gathersburg, MD: National Institute of Standards and Technology. November.

Buchanan, James M. 1987. "An Economic Theory of Clubs." In *Economics: Between a Predictive Science and Moral Philosophy.* College Station, TX: Texan A&M University Press, 1987

Caracostas, Paraskevas and Ugur Muldur. 2001. "The Emergence of the New European Union Research and Innovation Policy," in P. Laredo and P. Mustar, eds. *Research and Innovation Policies in the New Global Economy: An International Comparative Analysis.* Cheltenham, UK: Edward Elgar.

Chesbrough, Henry. 2003. *Open Innovation: The New Imperative for Creating and Profiting From Technology.* Cambridge, MA: Harvard Business School Press. April.

Cimoli, Mario and Marina della Giusta. 2000. "The Nature of Technological Change and its Main Implications on National and Local Systems of Innovation." IIASA Interim Report IR-98-029.

Coburn, Christopher and Dan Berglund. 1995. *Partnerships: A Compendium of State and Federal Cooperative Programs.* Columbus, OH: Battelle Press.

Combs, Kathryn L. and Albert N. Link. 2003. Innovation Policy in Search of an Economic Paradigm: The Case of Research Partnerships in the United States. *Technology Analysis & Strategic Management* 15(2).

Council on Competitiveness. 2005. *Innovate America: Thriving in a World of Challenge and Change.* Washington, D.C.: Council on Competitiveness.

Dahlman, Carl J. 2005. *India and the Knowledge Economy: Leveraging Strengths and Opportunities.* Washington, D.C.: World Bank Publications.

Dahlman, Carl J. and Jean Eric Aubert. 2001. *China and the Knowledge Economy: Seizing the 21st Century.* Washington, D.C.: The World Bank.

Dahlman, Carl and Anuja Utz. 2005. *India and the Knowledge Economy: Leveraging Strengths and Opportunities.* Washington, D.C.: The World Bank.

Daneke, Gregory A. 1998. "Beyond Schumpeter: Non-linear Economics and the Evolution of the U.S. Innovation System." *Journal of Socio-economics* 27(1):97-117.

De la Mothe and Gilles Paquet. 1998. "National Innovation Systems, 'Real Economies' and Instituted Processes." *Small Business Economics* 11:101-111.

Doloreux, David. 2004. "Regional Innovation Systems in Canada: A Comparative Study" *Regional Studies* 38(5):479-492.

Eaton, Jonathan, Eva Gutierrez, and Samuel Kortum. 1998. "European Technology Policy." NBER Working Papers 6827.

Edler, J. and S. Kuhlmann. 2005. "Towards One System? The European Research Area Initiative, the Integration of Research Systems and the Changing Leeway of National Policies." *Technikfolgenabschätzung: Theorie und Praxis*. 1(4):59-68.

Eickelpasch, Alexander and Michael Fritsch. 2005. "Contests for Cooperation: A New Approach in German Innovation Policy." *Research Policy* 34:1269-1282.

Endquist, Charles, ed. 1997. *Systems of Innovation: Technologies, Institutions, and Organizations*, London, UK: Pinter.

EOS Gallup Europe. 2004. *Entrepreneurship*. Flash Eurobarometer 146. January. Accessed at <*http://ec.europa.eu/enterprise/enterprise_policy/survey/eurobarometer146_en.pdf*>.

European Commission. 2003. "Innovation in Candidate Countries: Strengthening Industrial Performance." May.

Fangerberg, Jan. 2002. *Technology, Growth, and Competitiveness: Selected Essays*. Cheltenham, UK and Northampton, MA: Edward Elgar.

Federal Register Notice. 2004. "2004 WTO Dispute Settlement Proceeding Regarding China: Value-Added Tax on Integrated Circuits." April 21.

Feldman, Maryann and Albert N. Link. 2001. "Innovation Policy in the Knowledge-based Economy." *Economics of Science, Technology and Innovation Volume 23*. Boston, MA: Kluwer Academic Press.

Feldman, Maryann P., Albert N. Link, and Donald S. Siegel. 2002. The Economics of Science and Technology: An Overview of Initiatives to Foster Innovation, Entrepreneurship, and Economic Growth. Boston, MA: Kluwer Academic Press.

The Financial Times. 2006. "India Needs Big Infrastructure Drive." February 23.

Fonfria, Antonio, Carlos Diaz de la Guardia, and Isabel Alvarez. 2002. "The Role of Technology and Competitiveness Policies: A Technology Gap Approach." *Journal of Interdisciplinary Economics* 13(1-2-3):223-241.

Foray, Dominique and Patrick Llerena. 1996. "Information Structure and Coordination in Technology Policy: A Theoretical Model and Two Case Studies," *Journal of Evolutionary Economics*. 6(2).

Furman, Jeffrey L., Michael E. Porter, and Scott Stern. 2002. "The Determinants of National Innovative Capacity." *Research Policy* 31:899-933.

George, Gerard and Ganesh N. Prabhu. 2003. "Developmental Financial Institutions as Technology Policy Instruments: Implications for Innovation and Entrepreneurship in Emerging Economies." *Research Policy* 32(1):89-108.

Grande, Edgar. 2001. "The Erosion of State Capacity and European Innovation Policy: A Comparison of German and EU Information Technology Policies." *Research Policy* 30(6):905-921.

Hall, Bronwyn H. 2002. "The Assessment: Technology Policy." *Oxford Review of Economic Policy* 18(1):1-9.

Huang, Yasheng and Tarun Khanna. 2003. "Can India Overtake China?" *Foreign Policy* July–August.

Hughes, Kent. 2005. *Building the Next American Century: The Past and Future of American Economic Competitiveness*. Washington, D.C.: Woodrow Wilson Center Press. Chapter 14.

Hughes, Kent H. 2005. "Facing the Global Competitiveness Challenge." *Issues in Science and Technology*. XXI(4):72–78.

Jaffe, Adam B, Josh Lerner, and Scott Stern, eds. 2003. *Innovation Policy and the Economy: Volume 3*. Cambridge, MA: The MIT Press.

Jasanoff, Sheila, ed. 1997. *Comparative Science and Technology Policy*. Elgar Reference Collection. International Library of Comparative Pubic Policy, Volume 5. Cheltenham, UK and Lyme, NH: Edward Elgar.

Joy, William. 2000. "Why the Future Does Not Need Us." *Wired* 8.04. April.

Koschatzky, Knut. 2003. "The Regionalization of Innovation Policy: New Options for Regional Change?" In G. Fuchs and Phil Shapira, eds. *Rethinking Regional Innovation: Path Dependency or Regional Breakthrough?* London, UK: Kluwer, 2003.

Kuhlmann, Stephan and Jakob Edler. 2003. "Scenarios of Technology and Innovation Policies in Europe: Investigating Future Governance—Group of 3" *Technological Forecasting & Social Change* 70.

Lall, Sanjaya. 2002. "Linking FDI and Technology Development for Capacity Building and Strategic Competitiveness." *Transnational Corporations* 11(3):39–88.

Laredo, Philippe and Philippe Mustar, eds. 2001. *Research and Innovation Policies in the New Global Economy: An International Perspective.* Cheltenham, UK: Edward Elgar.

Lembke, Johan. *Competition for Technological Leadership: EU Policy for High Technology.* Cheltenham, UK and Northampton, MA: Edward Elgar, 2002.

Lemola, Tarmo. 2002. "Convergence of National Science and Technology Policies: The Case of Finland." *Research Policy* 31(8–9):1481–1490.

Lewis, James A. 2005. *Waiting for Sputnik: Basic Research and Strategic Competition.* Washington, D.C.: Center for Strategic and International Studies.

Lin, Otto. 1998. "Science and Technology Policy and its Influence on the Economic Development of Taiwan." In Henry S. Rowen, ed. *Behind East Asian Growth: The Political and Social Foundations of Prosperity.* London, UK and New York, NY: Routledge.

Maddison, Angus and Donald Johnston. 2001. *The World Economy: A Millennial Perspective.* Paris, France: Organization for Economic Cooperation and Development.

Mani, Sunil. 2004. "Government, Innovation and Technology Policy: An International Comparative Analysis." *International Journal of Technology and Globalization* 1(1).

McKibben, William. 2003. *Enough: Staying Human in an Engineered Age.* New York: Henry Holt & Co.

Meyer-Krahmer, Frieder. 2001. "The German Innovation System." Pp. 205–252 in P. Larédo, P. Mustar, eds. *Research and Innovation Policies in the New Global Economy: An International Comparative Analysis.* Cheltenham, UK: Edward Elgar.

Meyer-Krahmer, Frieder. 2001. Industrial Innovation and Sustainability—Conflicts and Coherence." Pp. 177–195 in Daniele Archibugi, Bengt–Ake Lundvall, eds. *The Globalizing Learning Economy.* NY: Oxford University Press.

Mody, Ashok and Carl Dahlman. 1992. "Performance and Potential of Information Technology: An International Perspective." *World Development* 20(12):1703-1719.

Murali, Kanta. 2003. "The IIT Story: Issues and Concerns." *Frontline* 20(3).

Mustar, Phillipe and Phillipe Laredo. 2002. "Innovation and Research Policy in France (1980–2000) or The Disappearance of the Colbertist State." *Research Policy* 31:55-72.

National Academy of Sciences/National Academy of Engineering/Insitute of Medicine. 2007. *Rising Above the Gathering Storm: Energizing and Employing America for a Brighter Economic Future.* Washington, D.C.: The National Academies Press.

National Research Council. 1996. Conflict and Cooperation in National Competition for High-technology Industry. Washington, D.C.: National Academy Press.

National Research Council. 1999. *The Advanced Technology Program: Challenges and Opportunities.* Charles W. Wessner, ed. Washington, D.C.: National Academy Press.

National Research Council. 1999. *Funding a Revolution: Government Support for Computing Research.* Washington, D.C.: National Academy Press.

National Research Council. 1999. *Industry-Laboratory Partnerships: A Review of the Sandia Science and Technology Park Initiative.* Charles W. Wessner, ed. Washington, D.C.: National Academy Press.

National Research Council. 1999. *New Vistas in Transatlantic Science and Technology Cooperation.* Charles W. Wessner, ed. Washington, D.C.: National Academy Press.

National Research Council. 1999. *The Small Business Innovation Research Program: Challenges and Opportunities.* Charles W. Wessner, ed. Washington, D.C.: National Academy Press.

National Research Council. 2000. *The Small Business Innovation Research Program: A Review of the Department of Defense Fast Track Initiative.* Charles W. Wessner, ed. Washington, D.C.: National Academy Press.

National Research Council. 2000. *U.S. Industry in 2000: Studies in Competitive Performance*. David C. Mowery, ed. Washington, D.C.: National Academy Press.

National Research Council. 2001. *The Advanced Technology Program: Assessing Outcomes*. Charles W. Wessner, ed. Washington, D.C.: National Academy Press.

National Research Council. 2001. *Building a Workforce for the Information Economy*. Washington, D.C.: National Academy Press.

National Research Council. 2001. *Capitalizing on New Needs and New Opportunities: Government-Industry Partnerships in Biotechnology and Information Technologies*. Charles W. Wessner, ed. Washington, D.C.: National Academy Press.

National Research Council. 2001. *A Review of the New Initiatives at the NASA Ames Research Center*. Charles W. Wessner, ed. Washington, D.C.: National Academy Press.

National Research Council. 2001. *Trends in Federal Support of Research and Graduate Education*. Stephen A. Merrill, ed. Washington, D.C.: National Academy Press.

National Research Council. 2002. *Government-Industry Partnerships for the Development of New Technologies: Summary Report*. Charles W. Wessner, ed. Washington, D.C.: The National Academies Press.

National Research Council. 2004. *The Small Business Innovation Research Program: Program Diversity and Assessment Challenges*. Charles W. Wessner, ed. Washington, D.C.: The National Academies Press.

National Research Council. 2005. *Getting Up to Speed: The Future of Superconducting*, Susan L. Graham, Marc Snir, and Cynthia A. Patterson, eds. Washington, D.C.: The National Academies Press.

National Research Council. 2007. *Enhancing Productivity Growth in the Information Age: Measuring and Sustaining the New Economy*. Dale W. Jorgenson and Charles W. Wessner, eds. Washington, D.C.: The National Academies Press.

National Research Council. 2007. *SBIR and the Phase III Challenge of Commercialization*. Charles W. Wessner, ed. Washington, D.C.: The National Academies Press.

Nelson, Richard R. and Katherine Nelson. 2002. "Technology, Institutions, and Innovation Systems." *Research Policy* 31:265-272.

Nelson, Richard R. and Nathan Rosenberg. 1993. "Technical Innovation and National Systems." In *National Innovation Systems: A Comparative Analysis*. Richard R. Nelson, ed. Oxford, UK: Oxford University Press.

Organisation for Economic Co-operation and Development. 1999. *Boosting Innovation: The Cluster Approach*. Paris, France: Organisation for Economic Co-operation and Development.

Organisation for Economic Co-operation and Development. 1999. *Managing National Innovation Systems*, Paris, France and Washington, D.C.: Organisation for Economic Co-operation and Development.

Organisation for Economic Co-operation and Development. 2001. *Social Sciences and Innovation*, Washington, D.C.: Organisation for Economic Co-operation and Development.

Organisation for Economic Co-operation and Development. 2004. "Summary Report: Micro-policies for Growth and Productivity." DSTI/IND(2004)7 Paris, France: Organization for Economic Cooperation and Development. October.

Oughton, Christine. 1997. "Competitiveness in the 1990s." *The Economic Journal* 107(444):1486-1503.

Oughton, Christine, Mikel Landabaso, and Kevin Morgan. 2002. "The Regional Innovation Paradox: Innovation Policy and Industrial Policy." *The Journal of Technology Transfer* 27(1).

Patel, P. and K. Pavitt. 1994. "National Innovation Systems: Why They are Important and How They Might be Compared?" *Economic Change and Industrial Innovation*.

Posen, Adam S. 2001. "Japan." In Benn Steil, David G. Victor, and Richard R. Nelson, eds. *Technological Innovation and Economic Performance*. Princeton, NJ: Princeton University Press.

President's Council of Advisors on Science and Technology. 2004. "Sustaining the Nation's Innovation System: Report on Information Technology Manufacturing and Competitiveness." Washington, D.C.: Executive Office of the President. January.

PricewaterhouseCoopers. 2006. "China's Impact on the Semiconductor Industry: 2005 Update." PricewaterhouseCoopers.

Rai, Saritha. 2006. "India Becoming a Crucial Cog in the Machine at I.B.M." *The New York Times*. June 5.

Reuters. 2006. "China Sees No Quick End to Economic Boom." February 21.

Romanainen, Jari. 2001. "The Cluster Approach in Finnish Technology Policy." Pp. 377–388 in Edward M. Bergman, Pim den Hertog, and David Charles, eds. *Innovative Clusters: Drivers of National Innovation Systems*. OECD Proceedings. Washington, D.C.: Organisation for Economic Co-operation and Development.

Ruttan, Vernon. 2002. *Technology, Growth and Development: An Induced Innovation Perspective*. Oxford, UK: Oxford University Press.

Rutten, Roel and Frans Boekema. 2005. "Innovation, Policy and Economic Growth: Theory and Cases." *European Planning Studies* 13(8).

Scherer, F. M. 2001. "U.S. Government Programs to Advance Technology." *Revue d'economie industrielle* 0(94):69-88.

Sheehan, Jerry and Andrew Wyckoff. 2003. "Targeting R&D: Economic and Policy Implications of Increasing R&D Spending." DSTI/DOC(2003)8. Paris, France: Organization for Economic Cooperation and Development.

Shin, Roy W. 1997. "Interactions of Science and Technology Policies in Creating a Competitive Industry: Korea's Electronics Industry." *Global Economic Review* 26(4):3-19.

Smits, Ruud and Stefan Kuhlmann. 2004. "The Rise of Systemic Instruments in Innovation Policy." *International Journal of Foresight and Innovation Policy*. 1(1/2).

Soete, Luc G. and Bastiann J. ter Weel. 1999. "Innovation, Knowledge Creation and Technology Policy: The Case of the Netherlands." *De Economist* 147(3). September.

Stanford University. 1999. *Inventions, Patents and Licensing: Research Policy Handbook*. Document 5.1. July 15.

Talele, Chitram J. 2003. "Science and Technology Policy in Germany, India and Pakistan." *Indian Journal of Economics and Business* 2(1):87-100.

Tansley, A. G. 1939. "British Ecology During the Past Quarter Century: The plant Community and the Ecosystem" *The Journal of Ecology* 27(2):513-530.

Tassey, Gregory. 2004. "Policy Issues for R&D Investment in a Knowledge-based Economy." *Journal of Technology Transfer* 29:153-185.

Teubal, Morris. 2002. What is the Systems Perspective to Innovation and Technology Policy and How can we Apply it to Developing and Newly Industrialized Economies? *Journal of Evolutionary Economics*. 12(1-2).

U.S. Department of Energy. 2006. Press Release. "Department Requests $4.1 Billion Investment as Part of the American Competitiveness Initiative: Funding to Support Basic Scientific Research." February 2.

U.S. General Accounting Office. 2002. *Export Controls: Rapid Advances in China's Semiconductor Industry Underscore need for Fundamental U.S. Policy Review*. GAO-020620. Washington, D.C.: U.S. General Accounting Office. April.

The Washington Post. 2006. "Chinese to Develop Sciences, Technology." February 10. P. A16.

Wessner, Charles W. 2005. "Entrepreneurship and the Innovation Ecosystem." In David B. Audretsch, Heike Grimm and Charles W. Wessner, eds. *Local Heroes in the Global Village: Globalization and the New Entrepreneurship Policies*. New York: Springer.

Yukio, Sato. 2001. "The Structure and Perspective of Science and Technology Policy in Japan." In Phillipe Laredo and Phillipe Mustar, eds. *Research and Innovation Policies in the New Global Economy: An International Comparative Analysis.* Cheltenham, UK and Northampton, MA: Edward Elgar, 2001.

Zeigler, Nicholas J. 1997. *Governing Ideas: Strategies for Innovation in France and Germany.* Ithaca, NY and London, UK: Cornell University Press.